The Open University

MT365 Graphs, networks and design

GW00482008

Design 1

Geometric design

Study guide

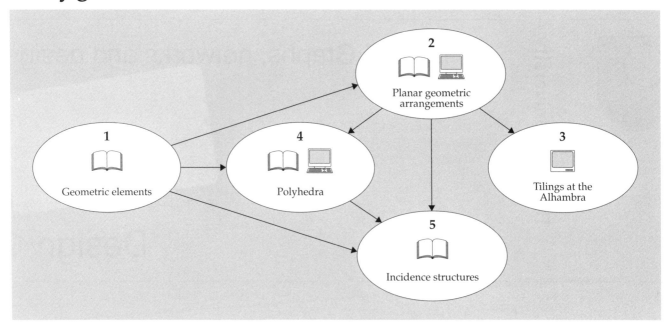

Section 1 gives the background to your study of geometry; you should find it straightforward. Section 2 will probably take up more of your time than any other section. Section 3 is a television section, in which examples of tilings which appear in the Alhambra at Granada in Spain are analysed. Section 4 is short — just a glimpse into the third dimension. Section 5 is rather like parts of Section 1 — more mathematical than visual; it is an important section, as it introduces concepts that are developed in the third and fourth *Design* units.

There are computer activities associated with Sections 2 and 4, which use the graphical capabilities of the computer to help you to visualize and manipulate tilings and polyhedra.

There is no audio-tape sequence associated with this unit.

The Open University, Walton Hall, Milton Keynes, MK7 6AA.

First published 1995. Reprinted 1998, 2002, 2003, 2005, 2008, 2009.

Printed and bound in Malta by Gutenberg Press Limited.

ISBN 0 7492 2216 6

1.5

Contents

Introduction

Geometry is a subject that makes an important contribution to the discipline known as *design*.

Many design problems are concerned with the shapes and arrangements of physical objects. For example, the design of a library or a wine cellar must take into account the efficient storage of its contents in such a way as to give easy access to any specified item. A study of the geometry of physical space is relevant to investigating such problems.

Geometry is concerned with the arrangement and juxtaposition of objects in a space. Often the objects are physical ones, such as tiles, honeycomb cells, the rooms in a building, or city blocks. The space can also be physical, such as the space within a room, the surface of the Earth, or the set of configurations that a desk lamp or a human body can assume.

Other design problems are more abstract. For example, if a number of drugs are to be tested on volunteers, but it is impracticable for every volunteer to test each drug, how would you design a suitable allocation of drugs to testers? For this problem, the objects are abstract entities — the pairings that represent these allocations of drugs to testers — and you need to design an experiment that is 'balanced', in the sense that the results must not be inadvertently biased by the chosen allocation of drugs to testers. Similarly, if you were asked to organize a three-round chess match between two teams of seven players each, how would you select who plays whom? Again, you want to find a solution that is 'balanced', in the sense that you want to ensure fair play and give all participants enjoyable games.

In both of these examples, you wish to make your allocations in a way that displays some of the sense of order, method and symmetry that was so beloved by the fictional Belgian detective Hercule Poirot. In effect, you are looking for a geometric structure — albeit one that does not arise directly out of any geometry of physical space.

In Section 1, *Geometric elements*, we introduce a geometry of physical space. We consider the concept of *dimension*, and look at the basic geometric elements in spaces of various dimensions. An important property of these elements is that they are *convex*. We explain what this means and show how to recognize and construct convex shapes and objects. We conclude with a description of some standard convex shapes and objects.

In Section 2, *Planar geometric arrangements*, we look at some arrangements of two-dimensional geometric elements in the plane — *tilings* and *polygonal animals*. In particular, we describe how the plane can be covered by regular shapes, such as regular polygons, six of which are shown below.

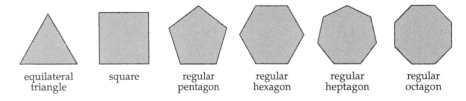

| equilateral triangle | square | regular pentagon | regular hexagon | regular heptagon | regular octagon |

Section 3, *Tilings at the Alhambra*, is a television section. The associated television programme extends some of the ideas in Section 2 by analysing some of the many tiling patterns to be found in the Alhambra at Granada in Spain.

Section 4, *Polyhedra*, is concerned with the three-dimensional analogues of polygons. We introduce *Euler's formula*, which relates the numbers of vertices, edges and faces of a convex polyhedron. We look at the way polyhedra can possess geometric regularity, and show that the five Platonic solids are the only regular convex polyhedra possible.

The Platonic solids were introduced in *Graphs 1*, Section 1.

| tetrahedron | octahedron | cube | icosahedron | dodecahedron |

In Section 5, *Incidence structures*, we investigate abstract geometric arrangements. These are used to solve problems like the drug-testing and chess-match allocation problems mentioned above.

Throughout this unit, angles are usually given in radians. The following table gives some angles in degrees with their radian equivalents.

degrees	0°	30°	45°	60°	90°	108°	120°	135°	150°	180°
radians	0	$\pi/6$	$\pi/4$	$\pi/3$	$\pi/2$	$3\pi/5$	$2\pi/3$	$3\pi/4$	$5\pi/6$	π

1 Geometric elements

In order to study the geometry of physical space, we first look at the concept of *dimension*. We then study some basic geometric elements, such as the vertices, edges and faces of a solid object, and investigate whether certain shapes and objects are *convex*. We conclude this section with a discussion of some standard examples of geometric shapes and objects.

1.1 Dimension

You are probably familiar with the notions that a line is *one-dimensional* and a plane *two-dimensional*, and with the description of the space we live in as *three-dimensional*. But exactly what is it about a line, a plane and physical space that justifies these descriptions?

Definition

The **dimension** of a space is the minimum number of coordinates that are needed to describe the position of a typical point in that space.

For example, a plane is two-dimensional, because only two coordinates are needed in order to describe the position of a typical point in it. One way of doing so is by means of *Cartesian coordinates*.

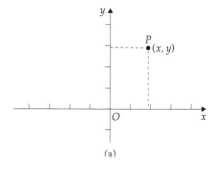

(a)

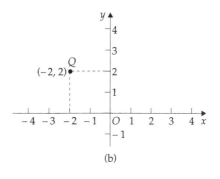

(b)

René Descartes (1596–1650) introduced the idea of specifying a point in a plane by two coordinates, although he was not responsible for the current description in terms of an origin and two rectangular axes.

We choose an origin O in the plane, and then choose two directions at right angles to serve as the x-axis and y-axis. The position of a typical point P in the plane is described by its x-coordinate and its y-coordinate; these coordinates are found by dropping perpendiculars from P to the respective axes, as indicated in diagram (a) above. In diagram (b), the x-coordinate of the particular point Q is –2 and the y-coordinate is 2. We describe the position of a typical point P by writing its x-coordinate and its y-coordinate as the *ordered pair* (x, y), so that Q is described by (–2, 2).

We shall not dwell on this, as we expect you to be familiar with Cartesian coordinates. Note, however, that several choices have to be made before the position of a point in a plane can be so described. First, we choose the origin. Then we choose the directions for the axes. Finally, we choose the scales for measuring distance along the axes.

Alternatively, we can describe the positions of points in a plane by using *polar coordinates*.

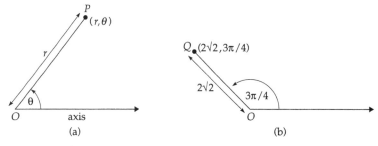

(a)   (b)

We again choose an origin O, but this time we choose just one axis, drawn outwards from O. We then describe the position of a typical point P by the ordered pair of numbers (r, θ), where r is the distance from O to P and θ ($0 \le \theta < 2\pi$) is the angle which the line OP makes with the axis, measured in an anticlockwise direction, as indicated in diagram (a) above. Thus, if we choose O to be the origin of the Cartesian system, and our axis to be the x-axis (in the direction of positive x), then the polar coordinates of our earlier point Q are (r, θ), where $r = 2\sqrt{2}$ and $\theta = 3\pi/4$, as shown in diagram (b).

Polar coordinates are so called because the origin is sometimes referred to as the *pole*.

Problem 1.1

(a) In the figure in the margin, which uses Cartesian coordinates, we have drawn a straight line L at an angle of $\pi/4$ to the x-axis and through the point (1, 0). What is the dimension of the space L?

(b) What is the dimension of the space comprising the single point P?

For two-dimensional spaces other than a plane, choices other than Cartesian or polar coordinates may be appropriate. For example, if you wish to describe the positions of points on the surface of a sphere, you can set up three rectangular axes and describe any point on the surface using the corresponding three coordinates. But this would be inefficient. Your system would actually be one for describing the three-dimensional space in which the sphere is embedded, and you would simply be treating the spherical surface as a subset of this space.

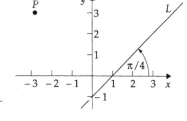

In fact, we can describe the position of any point on the spherical surface with just *two* coordinates, by selecting two great circles at right angles as an 'equator' and a 'Greenwich meridian', and describing the position of a typical point on the surface by its 'latitude' and 'longitude' with respect to these choices. Two is the minimum number of coordinates required, and so the surface of a sphere is two-dimensional.

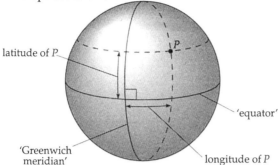

In *Design 2, Kinematic design*, you will meet more complicated physical spaces — for example, the set of possible positions and orientations for the components of a mechanical system, such as a robot. In the rest of this unit, however, the only physical spaces that you will meet are zero-dimensional points, one-dimensional lines, two-dimensional planes, and the three-dimensional space in which we live. These are called *Euclidean spaces*, because they satisfy the properties that were assumed by the ancient Greek geometer Euclid (*c.* 300 BC), such as:

(a) two distinct points lie on exactly one line;

(b) two distinct non-parallel lines meet in exactly one point.

The two Euclidean spaces with which we shall be particularly concerned are the two-dimensional Euclidean space represented by an infinite flat surface, which we shall refer to as *the plane*, and the three-dimensional Euclidean space represented by the infinite space in which we live, which we shall refer to as just *space*.

Many spaces, such as the surface of a sphere, are not Euclidean. For example, if you interpret 'lines' on a sphere as great circles, then any two of these 'lines' meet in two points rather than in one; also there are no parallel lines on this surface, in the Euclidean sense.

Note that, in the plane and in space, we shall use *line* to mean 'straight line'.

1.2 Basic geometric elements

There are many examples of systems whose primary function depends crucially on geometric design, ranging from wallpaper designs, via microelectronic circuits, to architectural and engineering structures. In each case, the system consists of many components, or building blocks, organized in a precise geometric arrangement. The physical nature of the components varies widely (ornamental motifs, light-emitting diodes, or girder beams, for example). However, within a particular system, there are often just one or two types of component. Similarly, the physical construction of the systems and the organization of their components can also vary widely, and yet it is often found that there are just a few arrangements that are optimal in some sense.

Here we are not concerned with the physical attributes or internal structure of the components, except for their shape and size. Our interest is to consider them geometrically, in terms of their vertices (corners), edges, surfaces and volumes.

Problem 1.2 —————————————————————————————————

Give two common examples, one natural and one man-made, of systems with only one type of component. Identify the component in each case.

We now investigate the basic geometric elements used to describe geometric components. In particular, we consider the questions: what are these basic elements? are they unique in some sense?

To answer these questions, we first examine a typical component, or building block, such as a brick. If we ignore its physical properties (hardness, weight, etc.), we can represent it adequately by its shape and size. Its shape has three common mathematical names: *cuboid, rectangular prism* and *rectangular parallelepiped*. The first name emphasizes that it can be obtained from a cube by changing the scale along one or more axes; the second, that it has rectangular cross-sections; and the third, that its opposite pairs of faces are rectangular and parallel.

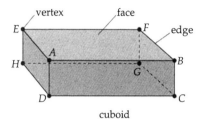

cuboid

Whatever you like to call it, we can describe its shape in terms of its *vertices, edges* and *faces*: its vertices are the *points A–H*; its edges are *line segments* joining pairs of points, such as *AB* or *EH*; and its faces are *plane segments* bounded by four line segments, such as *ABCD* (bounded by *AB, BC, CD* and *DA*). The cuboid itself is a *space segment* bounded by the plane segments.

By a *line segment* we mean a segment of a line bounded by two points, and by a *plane segment* we mean a segment of a plane bounded by (any number of) line segments.

Problem 1.3

Write down the dimensions of the vertices, edges and faces of a cuboid. What is the dimension of the cuboid itself?

The solution to Problem 1.3 is instructive, because it gives us an alternative way of thinking of the dimension of a *bounded* set of points. Our three-dimensional cuboid is bounded by six two-dimensional faces; each face is bounded by four one-dimensional edges; and each edge is bounded by two zero-dimensional points. Thus, a set of a given dimension is bounded by sets of one dimension less.

Note that each face can be defined by the edges bounding it; given any four edges that bound a face, there is only one possible face that they bound. However, we cannot just take any set of four line segments and expect them to bound a face. For example, the line segments *AB, BC, AE, EH* in the cuboid above do not bound a face. In fact, we are already in trouble with just the two line segments *AB* and *EH*, as no plane contains both of these line segments, and so they cannot be used to define a face of any object in space, let alone a cuboid. Thus, a set of line segments must have special properties in order to define a face — each line segment must share each of its bounding points with another line segment of the set, and the line segments must all lie in a common plane.

Note that each edge is determined uniquely by its two endpoints, as there is only one way to draw a line segment between two distinct points in space.

Similarly, the cuboid itself can be defined by the six faces bounding it — but again, we cannot take an arbitrary set of six plane segments and expect them to define a cuboid or any other object in space. They must 'fit together' along their bounding line segments.

Nevertheless, provided that we choose our points, line segments or plane segments appropriately, we have a situation in which an object (the cuboid), consisting of *infinitely many* points in space, can be defined by a *finite* number of basic geometric elements: eight points; or twelve line segments, each defined by two points; or six plane segments, each defined by four line segments.

The above discussion may suggest that a cuboid is a special case, and that only such 'angular' or 'squared' objects can be described in discrete terms in this way. However, this is not so — there is no reason why the angles between plane segments or line segments need be right angles. By allowing these angles to vary, we can use discrete sets of *basic geometric elements* (points, line segments, plane segments and space segments) to describe a wide range of shapes in the plane and objects in space. We can even approximate objects with curved edges and curved bounding surfaces (to any desired accuracy), by dividing the edges into a string of successive line segments and faceting the surfaces into an array of contiguous plane segments.

We use *shape* to mean a two-dimensional 'object' in the plane and restrict *object* to mean a three-dimensional 'object' in space.

With the advent of computer-aided design (CAD) systems and graphic-display terminals, this discretization of shapes in the plane and objects in space has become a very important design tool, because the shapes and objects can be conveniently represented by simpler data structures inside the computer, and displayed with simpler routines on the screen.

Without further justification, we adopt this discrete approach here and concentrate on shapes and objects constructed from basic geometric elements — points, line segments, plane segments and space segments. The following figure illustrates a point, a line segment, the simplest type of plane segment (a triangle) and the simplest type of space segment (a tetrahedron).

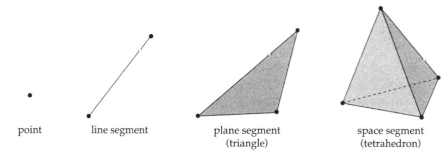

| point | line segment | plane segment (triangle) | space segment (tetrahedron) |

Note that the rectangular faces of a cuboid are not the simplest plane segments. Each face requires four points to define it; but if such a face is divided by a diagonal line, it can be regarded as being composed of two triangles, each of which requires just three points to define it.

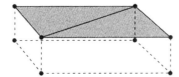

Similarly, a cuboid itself can be split up into tetrahedra, although this is more difficult to visualize. In the following figure, the three shaded planes split a cuboid into six tetrahedra.

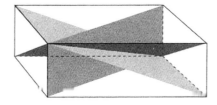

9

In order to state precisely what we mean by points *defining* line segments, triangles and tetrahedra, we need to consider the important concept of *convexity*; this is the subject of the next subsection.

1.3 Convexity

Intuitively, in space, an object is *convex* if its boundary has no depressions or inward bulges. Thus, in a convex lens, the surface of the glass bulges outwards, whereas in a concave lens it bulges inwards.

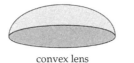 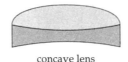

convex lens concave lens

Similarly, in the plane, an oval shape is convex, whereas a longitudinal cross-section of a banana is not.

convex shape non-convex shape

What about two-dimensional shapes whose boundaries are made up of line segments? The idea of 'bulging outwards' may not be applicable any more, but we can still distinguish between shapes that do not have indentations and those that do.

Similarly, some three-dimensional objects do not have indentations and some do.

 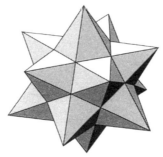

These intuitive ideas lead us to the following precise definition of a convex set of points.

Definition

A set S of points in a Euclidean space is **convex** if, for each possible choice of two points P and Q in S, all the points on the line segment joining P and Q also lie in S.

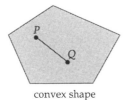 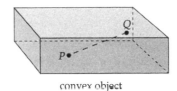

convex shape convex object

Note that it can be quite difficult to prove *convexity* directly from the definition, since we cannot always consider all possible points P and Q. However, proving *non-convexity* is usually easier, since it is sufficient to find *one* pair of points P and Q such that the line segment PQ does not lie wholly in the set, as shown in the following diagrams.

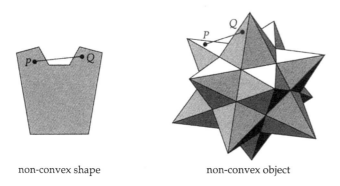

non-convex shape   non-convex object

Problem 1.4

Show that each of the following sets is non-convex by finding an appropriate line segment in each case.

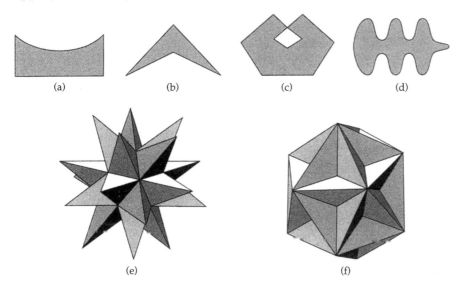

(a)   (b)   (c)   (d)

(e)   (f)

We can now say why two distinct points *define* the line segment between them, the three vertices of a triangle *define* the triangle, and the eight vertices of a cuboid *define* the cuboid. In each case, the shape or object so defined is the *smallest* convex set containing the points in question. Such a set is called the *convex hull* of the points.

Definition

Let S be a set of points in a Euclidean space. The **convex hull** of S is the smallest convex set containing S; it is denoted by $\langle S \rangle$.

It can be proved that such a smallest convex set always exists.

First, we consider the cases that can arise when we investigate the convex hull of a small set of points in the plane.

The convex hull of two distinct points

If P and Q are distinct points in the plane, then $\langle P, Q \rangle$ is the line segment PQ joining P and Q.

The convex hull of three distinct points

If P, Q, R are distinct points in the plane, then two cases can arise.

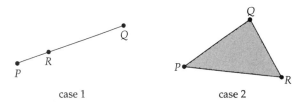

| case 1 | case 2 |

Case 1   If P, Q, R lie on a line, then one of them — R, say — lies between the other two, so $\langle P, Q, R \rangle$ is the line segment PQ.

Case 2   If P, Q, R do not lie on a line, then $\langle P, Q, R \rangle$ is the triangle PQR, with vertices P, Q, R.

The convex hull of four distinct points

If P, Q, R, S are distinct points in the plane, then four cases can arise.

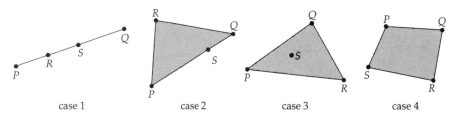

| case 1 | case 2 | case 3 | case 4 |

Case 1   If P, Q, R, S lie on a line, then two of them — R and S, say — lie between the other two, so $\langle P, Q, R, S \rangle$ is the line segment PQ.

Case 2   If exactly three of the points lie on a line — say, S lies between P and Q — but R does not lie on this line, then $\langle P, Q, R, S \rangle$ is the triangle PQR.

Case 3   If no three of P, Q, R, S lie on a line, but one of the points — S, say — lies within the triangle defined by the other three, then $\langle P, Q, R, S \rangle$ is the triangle PQR.

Case 4   If no three of P, Q, R, S lie on a line, and none of them lies within the triangle defined by the other three, then $\langle P, Q, R, S \rangle$ is the quadrilateral $PQRS$, with vertices P, Q, R, S.

Problem 1.5

The points P, Q, R, S have Cartesian coordinates $(0, 0)$, $(1, 0)$, $(0, 1)$ and $(\frac{1}{3}, \frac{1}{3})$, respectively. What shape is $\langle P, Q, R, S \rangle$?

How does your answer change if S has coordinates $(\frac{2}{3}, \frac{2}{3})$ or $(\frac{1}{2}, \frac{1}{2})$?

Example 1.1

Consider the six points P, Q, R, S, T, U in the figure shown.

The convex hull $\langle P, Q, R, S, T, U \rangle$ of these six points is the triangle PQR. This is because S, T, U all lie in the triangle defined by P, Q, R. In this case, the convex hull of the six points P, Q, R, S, T, U is the same as the convex hull of the three points P, Q, R; that is,

$$\langle P, Q, R, S, T, U \rangle = \langle P, Q, R \rangle.$$ ∎

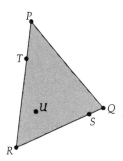

The general situation for points in the plane is described by the following theorem.

Theorem 1.1

Let S be a finite set of points in the plane that do not all lie on a line. Then the convex hull $\langle S \rangle$ is a convex polygon.

A **polygon** is a shape in the plane bounded by line segments (its *edges*). It consists of all the points on, and enclosed by, the bounding line segments.

Example 1.2

In the following figure, the eight points P, Q, R, S, T, U, V, W lie in the plane. The convex hull $\langle P, Q, R, S, T, U, V, W \rangle$ is the hexagon $PRSTUV$ and the convex hull $\langle P, Q, R, S, V, W \rangle$ is the quadrilateral $PRSV$.

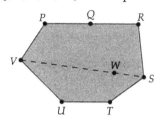

Problem 1.6

(a) In the figure in the margin, find four points from the set $\{P, Q, R, S, T, U\}$ whose convex hull is a quadrilateral, and four points whose convex hull is not a quadrilateral.

(b) What is the convex hull of the set $\{P, Q, S, T, U\}$?

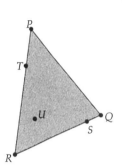

Let us now move on to consider the convex hull of a set of points in space.

Example 1.3

In the figure in the margin, the six points P, Q, R, S, T, U lie in space, but do not all lie in a plane. The points P, Q, R, S form the vertices of a tetrahedron; T lies on the edge PQ; and U lies on the face QRS. The convex hull $\langle P, Q, R, S, T, U \rangle$ is therefore the tetrahedron $PQRS$. ∎

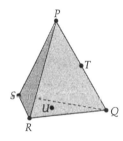

The analogue of Theorem 1.1 for points in space is as follows.

Theorem 1.2

Let S be a finite set of points in space that do not all lie in the same plane. Then the convex hull $\langle S \rangle$ is a convex polyhedron.

A **polyhedron** is an object in space bounded by plane segments (its *faces*). It consists of all the points on, and enclosed by, the bounding plane segments.

Example 1.4

In the following figure, the ten points $P, Q, R, S, T, U, V, W, X, Y$ lie in space, but do not all lie in a plane. The points P, Q, R, S, T, U, V, W form the vertices of a cuboid, and X and Y lie in the interior of the cuboid. The convex hull $\langle P, Q, R, S, T, U, V, W, X, Y \rangle$ is therefore the cuboid $PQRSTUVW$.

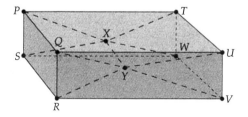

∎

For the figure in the margin, in which U lies on the face QRS, describe the convex hull of each of the following sets:

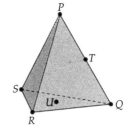

(a) $\{P, Q, R, T\}$;   (b) $\{P, T\}$;   (c) $\{P, Q, U\}$;

(d) $\{P, Q, R, U\}$;   (e) $\{P, Q, R, T, U\}$; (f) $\{P, Q, S, T\}$.

Convex hulls exist for n-dimensional Euclidean spaces, where $n > 3$. The n-dimensional analogue of a polygon or polyhedron is called a **polytope**, and the analysis of convex polytopes is a fascinating study in its own right. In this course, however, we restrict our attention to three dimensions at most, except for brief mentions of what the higher-dimensional analogues are and how they are formed.

Finally, in this subsection, note that the points and line segments bounding a convex hull can be interpreted as the vertices and edges of a graph, called the **graph** of the convex hull. For example, the graph of a rectangle is the cycle graph C_4 (which is the same as the 2-cube Q_2), and the graph of a cuboid is the 3-cube Q_3.

The graphs mentioned here were introduced in *Graphs 1*.

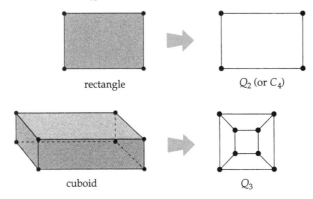

rectangle Q_2 (or C_4)

cuboid Q_3

1.4   Some standard convex shapes and objects

We conclude this section by considering some standard examples of convex geometric shapes and objects in Euclidean spaces of various dimensions.

Simplices

In each dimension, the 'simplest' type of convex set is called a *simplex*.

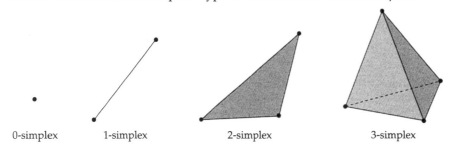

0-simplex   1-simplex   2-simplex   3-simplex

A single point is the *only* type of convex zero-dimensional set, and is called a **0-simplex**.

A line segment is the simplest type of convex one-dimensional set, and is called a **1-simplex**.

A triangle is the simplest type of convex two-dimensional set, and is called a **2-simplex**.

A tetrahedron is the simplest type of convex three-dimensional set, and is called a **3-simplex**.

Give examples of convex one-dimensional, two-dimensional and three-dimensional sets that are not simplices.

Simplices is the plural of *simplex*.

Note that a 0-, 1-, 2- or 3-simplex is the convex hull of its 1, 2, 3 or 4 defining (corner) points, which are placed in the most general possible positions in zero-, one-, two- or three-dimensional space respectively. Note also that any two of these defining points are joined by a line segment that forms part of the boundary of the simplex.

These observations lead us to deduce that an **n-simplex**, the simplest type of convex n-dimensional set, is the convex hull of a set of $n + 1$ points placed in the most general possible positions in n-dimensional space. Each line segment between pairs of these points bounds the n-simplex.

Problem 1.9

Draw the graphs of the 0-, 1-, 2- and 3-simplices. What is the graph of an n-simplex?

Orthotopes

Triangles are all very well, but in some respects a square or a rectangle is simpler. A square has equal sides, opposite sides are parallel, and its angles are right angles. A rectangle also has right angles and parallel pairs of sides, although the two pairs may differ in length. On the other hand, no two sides of a triangle can be parallel, and at most one angle of a triangle can be a right angle. Also, we would not normally consider splitting the faces of a cuboid into triangles, or the whole cuboid into tetrahedra, in order to get a better visualization or understanding of its geometry!

In a sense, this is a cultural rather than a mathematical preference. Our culture uses right angles in many situations where other angles would be possible. Walls are usually at right angles to the horizontal, as this is the easiest way to give them stability, but there is no fundamental reason why the floor plan of a building must be rectangular. Many cultures have non-rectilinear buildings, and indeed architects do sometimes experiment with non-rectilinearity — but even the most avant-garde of modern artists normally paint their works within rectangular borders! The following pictures illustrate some non-rectilinear buildings.

There may well be good reasons for this preference for right angles. They are easy to construct and recognize, possibly because perceiving when one is at right angles to the horizontal is fundamental to a species that normally balances on two legs! Be that as it may, this is a course in mathematics rather than anthropology, and our culture finds right angles easier to come to terms with than other angles, so it is reasonable to consider simple geometric shapes and objects that involve only right angles.

In two dimensions, the shapes that involve only right angles are rectangles, with squares as a special case. In three dimensions, the objects are cuboids, with cubes as a special case. The analogues of rectangles and cuboids in higher dimensions are called **orthotopes**.

To see how the orthotopes are formed, we begin with a point and translate it in a straight line from its initial position to a new position, so that the point traces out a line segment. If the line segment is now translated to a new position along a direction at right angles to the original translation of the point, then it traces out a rectangle. If the rectangle is now translated to a new position along a direction at right angles to both of the previous translation directions, the result is a cuboid.

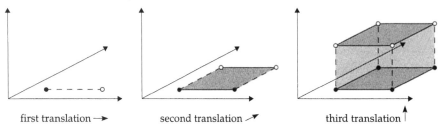

first translation $\longrightarrow$ second translation $\nearrow$ third translation $\uparrow$

In an n-dimensional space, it is possible to find n directions at right angles to each other, and so we can form an n-**orthotope** by continuing this process until all n of these directions have been used.

Problem 1.10 ———————————————————————

How many vertices (corners) does an n-orthotope have?

_____

We construct the **standard n-orthotope O_n** as follows:

O_1 is the *unit interval* with endpoints $x = 0$ and $x = 1$;

O_2 is the *unit square* formed by moving O_1 from $y = 0$ to $y = 1$, so that the vertices are at the points $(0, 0)$, $(1, 0)$, $(0, 1)$, $(1, 1)$;

O_3 is the *unit cube* formed by moving O_2 from $z = 0$ to $z = 1$, so that the vertices are at the points

$$(0, 0, 0), (1, 0, 0), (0, 1, 0), (1, 1, 0), (0, 0, 1), (1, 0, 1), (0, 1, 1), (1, 1, 1).$$

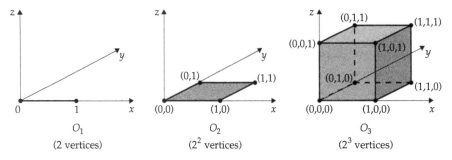

O_1 O_2 O_3
(2 vertices)   (2^2 vertices)   (2^3 vertices)

In general, the standard n-orthotope O_n is formed from O_{n-1} by adding an nth coordinate with value 0 or 1 to each of the 2^{n-1} points forming the vertices of O_{n-1}, so that O_n has $2 \times 2^{n-1} = 2^n$ vertices. In other words, the vertices of O_n are at all 2^n possible points with n coordinates each of which is 0 or 1.

We deduce that the graph of the standard n-orthotope is the n-cube Q_n.

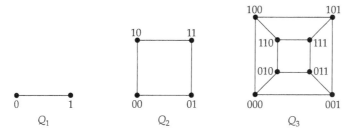

Q_1 Q_2 Q_3

Problem 1.11

How many edges does the graph of an n-orthotope have?

Hint   Use the handshaking lemma from *Graphs 1*.

Cross polytopes

A **cross polytope** is a convex polytope formed by taking the convex hull of the 'cross' formed by going out a certain distance in each direction (positive and negative) along each coordinate axis. We construct a **standard cross polytope** by going out a distance 1 along each axis.

For example, in one-dimensional space, we form the 'cross' whose endpoints are at $x = -1$ and $x = 1$; in this case the 'cross' is just a line segment. Then we form the convex hull of the 'cross', which is just the line segment again.

'cross'   standard one-dimensional
cross polytope

In two-dimensional space, we form the cross whose endpoints are at $(1, 0)$ This time it really is a cross. and $(-1, 0)$ and at $(0, 1)$ and $(0, -1)$. Then we form the convex hull of the cross, which is the same as the convex hull of the points themselves. It is a square.

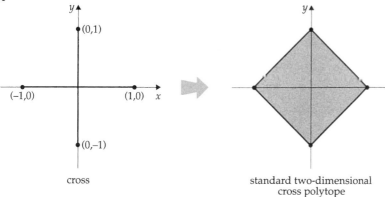

cross   standard two-dimensional
cross polytope

In three-dimensional space, we form the 'cross' whose endpoints are at $(1, 0, 0)$ and $(-1, 0, 0)$, at $(0, 1, 0)$ and $(0, -1, 0)$, and at $(0, 0, 1)$ and $(0, 0, -1)$. Then we form the convex hull of the 'cross', which again is the same as the convex hull of the points themselves. It is a regular octahedron.

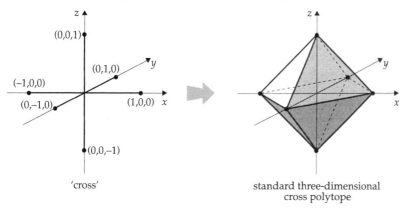

'cross'   standard three-dimensional
cross polytope

17

In n-dimensional space, we get the **standard n-dimensional cross polytope**.

The graph of a cross polytope is formed by drawing an edge between each pair of points except those on the same coordinate axis. In the one-dimensional case we get just two vertices and no edges, so the graph is the null graph N_2. In the two-dimensional case we get a square with four vertices each of degree 2, so the graph is the cycle graph C_4. In the three-dimensional case we get an octahedron with six vertices each of degree 4, which we can draw as our standard octahedron graph.

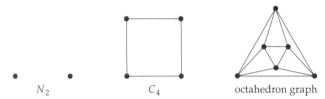

N_2 C_4 octahedron graph

The graph of an n-dimensional cross polytope is sometimes called an **n-octahedron**.

Problem 1.12 ────────────────────────────────

How many vertices, and how many edges, has an n-octahedron?

Hint   Use the handshaking lemma from *Graphs 1*.

Spheres

Finally, in each dimension, we can define a convex set which is the set of all points up to a given distance away from a given point. Such a set is called an **n-sphere**.

For example, a 1-sphere is a line segment, since a line segment of length $2a$ is the set of points whose distance from the midpoint is at most a.

A 2-sphere is a *disc*, and a 3-sphere is a solid *sphere* in the usual sense of the word.

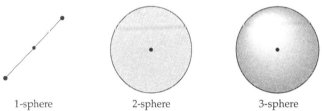

1-sphere   2-sphere   3-sphere

Spheres are very natural geometric objects, but they are different from our other standard examples of convex shapes and objects in that, except in one dimension, there is no finite set of points that has a sphere as its convex hull.

After studying this section, you should be able to:

- understand the concept of the *dimension* of a geometric space;

- describe the simplest basic geometric elements (*points, line segments, triangles* and *tetrahedra*) in Euclidean spaces of dimension up to three;

- understand the concept of a *convex set* and recognize such a set;

- construct the *convex hull* of a finite set of points in Euclidean spaces of dimension up to three;

- explain the terms *simplex, orthotope, cross polytope* and *sphere*.

---

2 Planar geometric arrangements

In Section 1 we dealt with the simpler types of basic geometric element (or building block) from which geometric shapes and objects are constructed. We now consider how these elements can be organized in the plane.

2.1 Packing bricks and discs

A brick wall is usually constructed from several copies of a single element — the brick (an orthotope) — arranged in a regular repeating two-dimensional pattern:

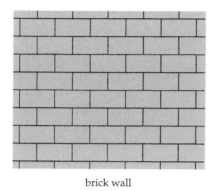

brick wall

If we consider just the positions of the bricks, rather than their shapes, then we can represent the position of each brick by a dot placed at its centre.

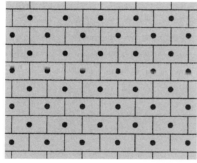

dots at centres of bricks

This gives a regular array of points, indicating the relative positions of the bricks. Such an array of points is called a **point lattice**.

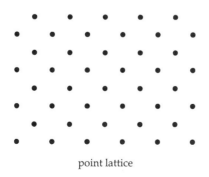

point lattice

If we now join the dots corresponding to adjacent bricks, we obtain the following pattern of triangles.

We consider two bricks to be *adjacent* if they have edges or parts of edges adjacent; having corners adjacent is not sufficient.

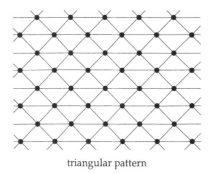

triangular pattern

Before the wall was built, the bricks may have been stacked together in a somewhat different arrangement. Again, we can represent their positions with dots.

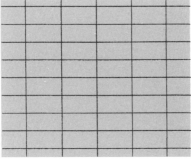

stack of bricks

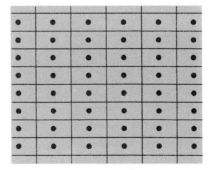

dots at centres of bricks

This time we get a different point lattice. And, joining the dots corresponding to adjacent bricks, we obtain a pattern of rectangles.

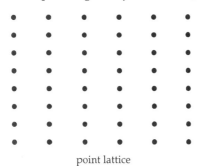

point lattice

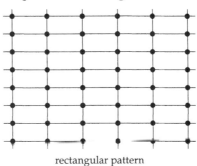

rectangular pattern

Each such arrangement of bricks or other objects in space that gives rise to a two-dimensional pattern is referred to as a **packing**.

Problem 2.1 ————————————————————————

For each of the following packings of square and triangular bricks, construct the corresponding point lattice, and illustrate the result of joining the dots corresponding to adjacent bricks.

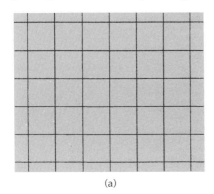

(a)

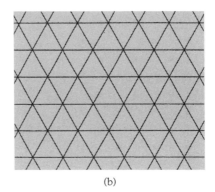

(b)

Because some packings of objects in space give rise to two-dimensional patterns, it is natural also to consider packings of shapes in the plane. For example, given a set of discs of the same radius, we can pack them in the plane as shown below. We obtain the corresponding point lattice by placing a dot at the centre of each disc and joining the dots corresponding to touching discs. This gives the square point lattice and the pattern of squares that you met in the above problem. Such an arrangement of discs is called a *square packing of discs*. In fact, any packing of shapes in the plane or objects in space that gives rise to a square point lattice is known as a *square packing arrangement*.

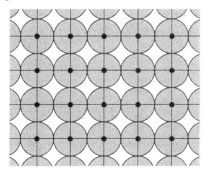
square packing of discs

The square packing of discs is a highly symmetrical arrangement, but it is not the most efficient packing of discs, in terms of minimizing wasted space. For, if the discs all have radius 1, then each square in the diagram above has area 4 and covers four quarter-discs, of total area π, and so the proportion of wasted space is $(4 - \pi)/4$, or about 21%.

However, consider the following packing arrangement:

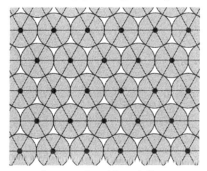
hexagonal packing of discs

Each disc now touches six neighbouring discs. The corresponding point lattice gives a pattern of equilateral triangles. There is still some wasted space, but not as much as in the square packing arrangement; in fact, since each triangle has area $\sqrt{3}$ and covers three disc sectors, each consisting of one-sixth of a disc, the total area of disc covered by each triangle is $\pi/2$, and so the proportion of wasted space is now $(\sqrt{3} - \pi/2)/\sqrt{3}$, or about 9%.

Such a packing of discs gives rise to the same triangular point lattice and the same pattern of equilateral triangles as does a packing of regular hexagons in the plane without gaps, as shown below, and is called a *hexagonal packing of discs*. In fact, any packing that gives rise to a triangular point lattice is known as a *hexagonal packing arrangement*.

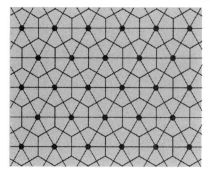
hexagonal packing arrangement

The hexagonal packing arrangement is very common in nature, and occurs in systems such as the wax cells of a bees' honeycomb, the arrangement of light-sensitive cells on the retina, and the convection cells in the Earth's atmosphere or on the surface of the Sun. It also occurs in manufactured systems such as aircraft wings, which are often constructed as a sandwich structure of two outer sheet-metal skins enclosing an inner honeycomb filling.

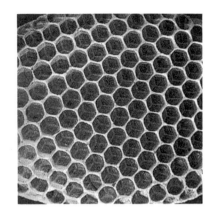

We have seen that discs cannot be packed together in the plane without some wasted space, but that squares, regular hexagons and equilateral triangles can be packed in regular arrangements without wasted space. We now investigate infinite planar 'packings without wasted space', otherwise known as *tilings*.

2.2 Tilings with polygons

Tilings are constructed from **tiles**, by which we mean planar shapes of finite area with no holes or gaps in them.

Definition

A **tiling** of the plane is a covering of the whole plane with tiles in such a way that there are no gaps or overlaps.

A tiling of the plane is often called a **tessellation**.

This is a very broad definition; all the drawings (a)–(i) are tilings.

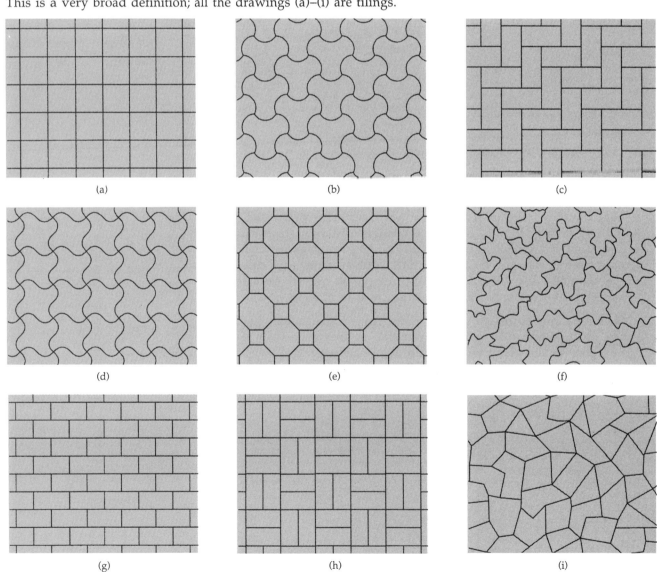

(a) (b) (c)

(d) (e) (f)

(g) (h) (i)

Note that the tiles need not be congruent; for example, each of tilings (e), (f) and (i) has tiles with more than one shape. Nor do the tiles need to be straight-edged, as tilings (b), (d) and (f) illustrate. Nor do they need to be convex, as tilings (b), (d), (f) and (i) illustrate.

Two planar shapes are *congruent* if they are the same size and shape — that is, if one can be superimposed exactly on the other, possibly after turning it over.

Many examples of tilings occur in the Alhambra at Granada in Spain, and in the work of the Dutch artist Maurits Escher.

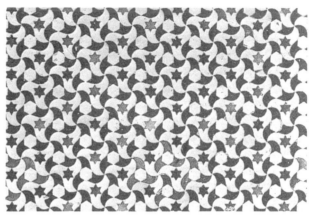

The study of tilings is a fascinating subject. However, here we can do no more than introduce the subject. We therefore restrict our study to tilings with straight-edged tiles — that is, *polygonal tilings*. So we exclude tilings like (b), (d) and (f) above.

In fact, we restrict our attention still further. Consider the 'brick wall' tiling (g) above. We can obtain infinitely many variations on this tiling by sliding the horizontal rows of bricks relative to one another through all possible distances up to the length of one brick. We shall exclude such possibilities by insisting that our polygons meet in such a way that each edge of each polygon corresponds exactly with an edge of another, and that no two polygons meet along more than one edge. This restricts the number of distinct polygonal tilings possible. We therefore make the following definition.

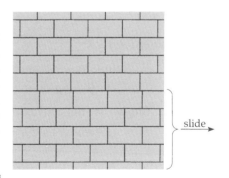

Definition

A polygonal tiling is an **edge-to-edge tiling** if each edge of each polygon is an edge of another polygon, and no two polygons meet along more than one edge.

For example, of the polygonal tilings above — namely tilings (a), (c), (e), (g), (h) and (i) — only tilings (a) and (e) are edge-to-edge tilings. Tilings (c), (g) and (h) fail to be edge-to-edge because it is not true that *each* edge of each polygon is an edge of another; tiling (i) fails because some polygons meet along more than one edge.

Which of the following are edge-to-edge tilings?

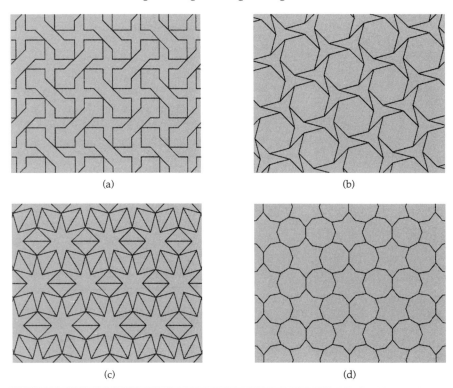

(a)

(b)

(c)

(d)

We shall restrict our attention to edge-to-edge tilings.

A further, natural, restriction arises from our discussions in Section 1, where we saw that the basic geometric elements used to build up geometric systems are convex hulls of sets of points. In the plane, these convex hulls are the convex polygons. We shall therefore restrict our attention to *edge-to-edge tilings using only convex polygons*, so that all the tilings in Problem 2.2 are excluded.

Even so, the subject is still a vast one, with many unsolved problems. For example, it is known that certain convex pentagons tile the plane, whereas certain others (such as the regular pentagon) do not. It is not currently known how to recognize the class of *all* convex pentagons that tile the plane.

A given shape *tiles the plane* if the plane can be tiled using only congruent copies of that shape in an edge-to-edge tiling.

Problem 2.3

(a) Explain why we cannot tile the plane with regular pentagons.

   Hint Each interior angle of a regular pentagon is $3\pi/5$ (= 108°).

(b) Show that the convex pentagon on the right tiles the plane.

   Hint Consider how to make rows of these pentagons.

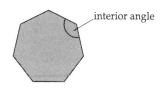

Finally, to ensure that our study of tilings is manageable, we further restrict our attention not just to edge-to-edge tilings with convex polygons but to edge-to-edge tilings with *regular* polygons. This enables us to establish a number of results using straightforward mathematical reasoning.

2.3   Tilings with regular polygons

Recall that a *regular polygon* is a polygon in which the edges are all of equal length, and the interior angles between each pair of adjacent edges are all equal. A regular polygon is necessarily convex, but a convex polygon need not be regular.

Derive an expression for the interior angle of an n-sided regular polygon.

Hint The sum of the interior angles of a triangle is π.

We supplement the general formula that you have just derived with the following table, which gives the interior angle θ (in radians and degrees) of an n-sided regular polygon for $n = 3, \ldots, 12$.

n	3	4	5	6	7	8	9	10	11	12
θ	$\pi/3$	$\pi/2$	$3\pi/5$	$2\pi/3$	$5\pi/7$	$3\pi/4$	$7\pi/9$	$4\pi/5$	$9\pi/11$	$5\pi/6$
	$60°$	$90°$	$108°$	$120°$	$129°$	$135°$	$140°$	$144°$	$147°$	$150°$

Some of the angles in the bottom row are given to the nearest degree. A table giving some other radian/degree equivalents appears in the Introduction to this unit.

Regular tilings

In any edge-to-edge tiling using regular polygons, the tiles fit together in such a way that each edge belongs to just two tiles and each vertex belongs to at least three tiles. We therefore have at least three edges radiating from each vertex, and at least three polygons arranged around each vertex. What arrangements of regular polygons are possible around a given vertex in an edge-to-edge tiling?

If we require all the regular polygons to be congruent, then the answer is not difficult to find, and gives us a *regular tiling*.

Definition

A **regular tiling** is an edge-to-edge tiling with congruent regular polygons.

The sum of the angles around any vertex is 2π, and so 2π must be an integer multiple of the interior angle θ. There are only three possibilities, namely $\theta = \pi/3$, $\pi/2$ and $2\pi/3$. These give rise to the following **vertex types** — that is, arrangements of congruent regular polygons around a vertex:

6 equilateral triangles	4 squares	3 regular hexagons

The integers represent the numbers of sides in the polygons: in the first case, there are six equilateral triangles meeting at each vertex; in the second case, there are four squares; and, in the third case, there are three regular hexagons. These arrangements give rise to the following regular tilings:

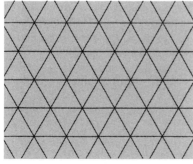
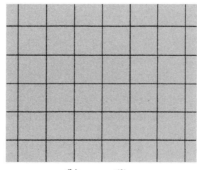
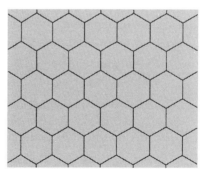

(a) triangular tiling	(b) square tiling	(c) hexagonal tiling

These three are the only regular tilings. It is convenient to introduce a 'code' for referring to tilings without drawing them. This is done in terms of the vertex types present — we merely write the sequence of integers representing the polygons encountered in circling around a vertex, and separate them by dots. The triangular tiling has the code 3.3.3.3.3.3, as there are six triangles around each vertex. Similarly, the square tiling has the code 4.4.4.4, and the hexagonal tiling has the code 6.6.6.

Semi-regular tilings

The situation is more complicated if two or more types of regular polygon are present at each vertex. In this case there are eighteen possible vertex types; these and their codes are illustrated below.

Note that we count mirror images such as the following as being the same; this is consistent with regarding shapes that are mirror images as congruent.

For the codes, our convention is that it does not matter where we start in circling around each vertex, so we regard 4.6.12 as equivalent to 6.12.4 or 12.4.6; also, we may circle around each vertex clockwise or anticlockwise, so 4.12.6 is also equivalent to 4.6.12.

We have now examined the possible arrangements of two or more polygons around a vertex. What tilings arise from these vertex types?

Definition

A **semi-regular tiling** is an edge-to-edge tiling with regular polygons, not all congruent, such that each vertex type is the same.

There are eight semi-regular tilings, corresponding to the first eight vertex types shown above. These are shown at the top of the next page.

The first of these comes in two distinct 'mirror image' forms, as shown in the margin; but as we count mirror-image vertex types as the same, we must do the same with types of tiling.

You may wonder what happened to the other ten vertex types — why do they not lead to semi-regular tilings? The answer is that, if you try to build up a tiling using any one of them, the pattern quickly breaks down.

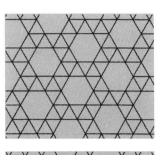

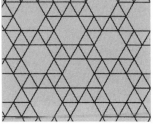

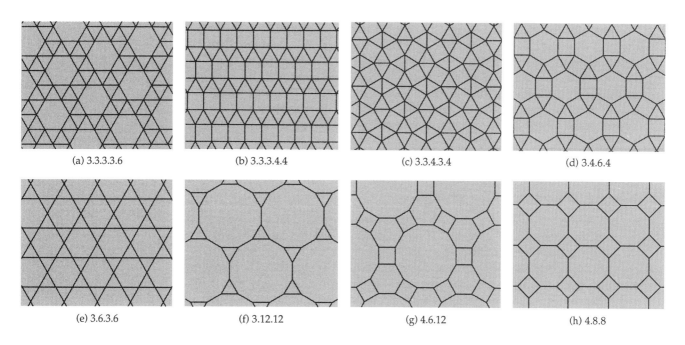

(a) 3.3.3.3.6 (b) 3.3.3.4.4 (c) 3.3.4.3.4 (d) 3.4.6.4

(e) 3.6.3.6 (f) 3.12.12 (g) 4.6.12 (h) 4.8.8

Demi-regular tilings

Some of the other vertex types can be used, however, provided that we drop the requirement that all the vertex types in a tiling be the same. This gives us the *demi-regular tilings*.

> **Definition**
>
> A **demi-regular tiling** is an edge-to-edge tiling with regular polygons in which more than one vertex type occurs.

It follows that, in a demi-regular tiling, not all the vertices have the same arrangement of polygons around them, and so there must be more than one type of regular polygon present. There are infinitely many demi-regular tilings; however, only twelve of the eighteen possible vertex types shown above — namely, the first thirteen with the exception of 5.5.10 — plus the three regular vertex types — 3.3.3.3.3.3, 4.4.4.4 and 6.6.6 — can occur in such a tiling. Some demi-regular tilings are shown below.

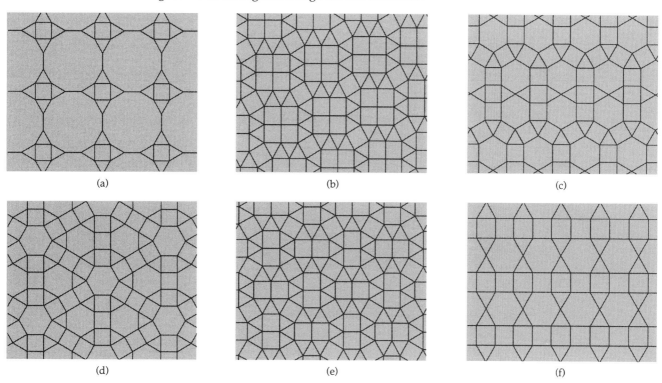

(a) (b) (c)

(d) (e) (f)

The fact that this class of tilings is infinite is easily demonstrated by noticing that the demi-regular tiling (f) has 'fault lines' — namely, the horizontal lines bounding the rows of square tiles. If we take one of these fault lines and shift all the tiles above it by (say) one edge length to the left relative to the tiles below it, then we get a new demi-regular tiling, as shown in the margin. By repeating this operation on different rows, we can create infinitely many variations on the original tiling.

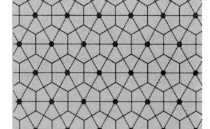

fault line

Another reason why there are infinitely many demi-regular tilings is that, in a tiling containing regular hexagons, any hexagon can be split into six equilateral triangles. Since these hexagons can be chosen at random, there are infinitely many possibilities.

We have seen that we can define a regular or semi-regular tiling uniquely by specifying the code of its vertex type. To describe a demi-regular tiling, we list the codes corresponding to all the vertex types that are present; for example, the code for the demi-regular tiling (a) is 3.4.3.12/3.12.12. Here, the code does not always define the demi-regular tiling uniquely, since, for example, arrangements with fault lines can give rise to many different tilings with the same code.

Problem 2.5

List the codes for each of the demi-regular tilings (b)–(f).

2.4 Dual tilings

At the beginning of this section, we illustrated some packings of the plane with various shapes. In order to illustrate the adjacencies between neighbouring shapes, we formed the point lattice by placing dots at the centres of the shapes. We then joined the dots corresponding to adjacent shapes, as illustrated on the right.

We can carry out a similar operation for any edge-to-edge tiling with regular polygons, giving the so-called *dual tiling*. This process essentially replaces each tile with a vertex, and each vertex with a tile.

You will encounter this 'role reversal' process again in Sections 4 and 5, and also in *Graphs 3, Planarity and colouring*, in *Design 3, Design of codes*, and in *Design 4, Block designs*. The general name in mathematics for such 'role reversal' is *duality*.

Edge-adjacent polygons are polygons that share a common edge.

> ### Definition
>
> Given an edge-to-edge tiling with regular polygons, we construct its **dual tiling** by joining the centre of each polygon to the centres of its edge-adjacent polygons, using line segments.

The construction of the dual of the triangular tiling is illustrated below.

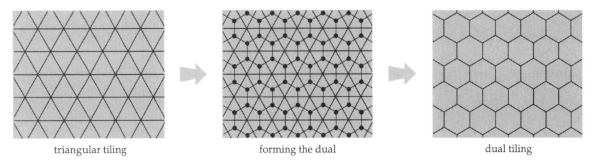

triangular tiling   forming the dual   dual tiling

The duals of the regular and semi-regular tilings are as follows.

Duals of the regular tilings

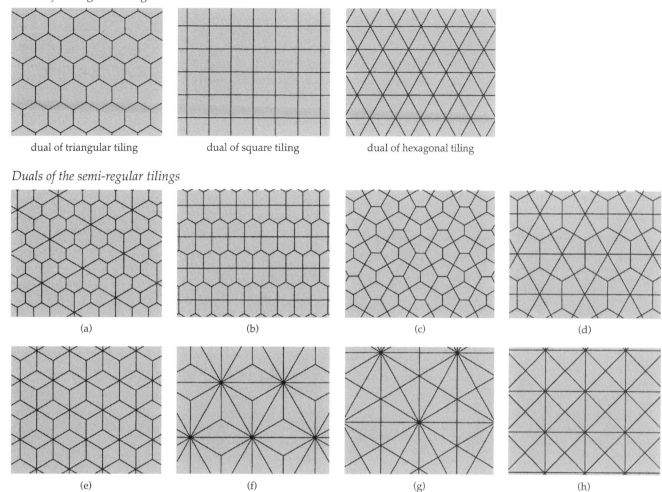

dual of triangular tiling   dual of square tiling   dual of hexagonal tiling

Duals of the semi-regular tilings

(a)   (b)   (c)   (d)

(e)   (f)   (g)   (h)

Problem 2.6

Do any regular or semi-regular tilings have duals that are themselves regular or semi-regular tilings? If so, which ones?

The duals of the regular and semi-regular tilings are called the **Laves tilings**, after the early twentieth-century crystallographer Fritz Laves. As you saw in Problem 2.6, three of them are regular tilings — the triangular and hexagonal tilings are duals of each other, and the square tiling is self-dual. The duals of the eight semi-regular tilings are not regular, semi-regular or demi-regular, as they do not contain regular polygons.

Problem 2.7

The following figure shows the two packings of bricks that we considered at the beginning of the section. Sketch the dual tilings of these tilings.

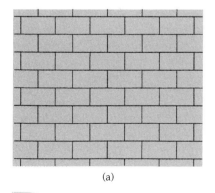

(a)

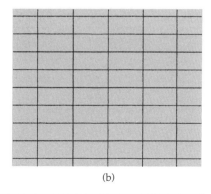

(b)

2.5 Polygonal animals

The tilings we considered above are *infinite* patterns of polygons in the plane. We now examine some *finite* patterns, and restrict our attention to those arrangements constructed from a finite number of regular polygons in *edge-to-edge contact* — that is, where each common edge is shared exactly by two polygons.

> **Definition**
>
> A **polygonal animal** is a connected arrangement of non-overlapping regular polygons in edge-to-edge contact. If it comprises exactly *n* polygons, it is called an *n*-**animal**.

Some examples of 5-animals are shown below.

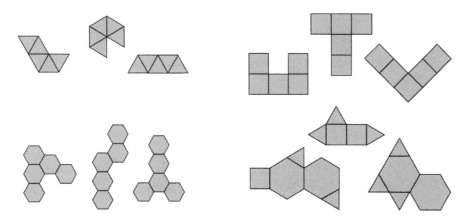

Of particular interest are the polygonal animals formed from equilateral triangles only, from squares only, and from regular hexagons only, because each can be regarded as a finite segment of the corresponding regular tiling. These have special names, as follows.

> **Definition**
>
> An *n*-animal formed from equilateral triangles is called an *n*-**iamond**.
>
> An *n*-animal formed from squares is called an *n*-**omino**.
>
> An *n*-animal formed from regular hexagons is called an *n*-**hex**.

The names *n-iamond* and *n-omino* are generalizations of the words *diamond* and *domino*, respectively.

We regard two polygonal animals as the *same* if one can be superimposed exactly on the other, possibly after turning it over, in such a way that each polygon of one fits exactly over a polygon of the other. Thus, for example, we regard the following 4-ominoes as being the same:

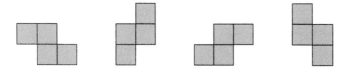

Problem 2.8

Draw all the distinct *n*-iamonds and *n*-ominoes for *n* = 1, 2, 3, 4 and all the distinct *n*-hexes for *n* = 1, 2, 3.

You have seen the *n*-ominoes for *n* = 1, 2, 3, 4 in the *Introduction* unit, Exercise 2.6.

It is an unsolved problem to determine how many distinct *n*-iamonds, *n*-ominoes and *n*-hexes can be formed for a general value of *n*.

The **graph** of a polygonal animal is formed in a similar manner to the dual of a tiling. We place vertices in the centres of the polygons and join the vertices corresponding to edge-adjacent polygons.

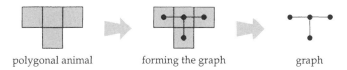

polygonal animal   forming the graph   graph

Problem 2.9

Draw the graphs of all the *n*-iamonds and *n*-ominoes for *n* = 1, 2, 3, 4 and of all the *n*-hexes for *n* = 1, 2, 3. How many non-isomorphic graphs are there in each case?

2.6 Computer activities

The computer activities for this section are described in the *Computer Activities Booklet*.

> After studying this section, you should be able to:
>
> - describe the square and hexagonal packing arrangements of discs in the plane;
> - explain the terms *tiling*, *polygonal tiling*, and *edge-to-edge tiling*;
> - determine which regular polygons can be arranged to fit around a vertex in the plane such that they are in edge-to-edge contact;
> - distinguish between *regular*, *semi-regular* and *demi-regular tilings*;
> - recognize the three regular and the eight semi-regular tilings, and describe the method of coding used for describing these;
> - explain what is meant by the *dual* of an edge-to-edge tiling with regular polygons;
> - recognize the eleven *Laves tilings* and understand the duality relationship between them and the regular and semi-regular tilings;
> - explain the terms *polygonal animal*, *n-omino*, *n-iamond*, *n-hex* and the *graph* of a polynomial animal, and give examples for small values of *n*.

3   Tilings at the Alhambra

This section comprises the television programme related to this unit, on tiling patterns to be found in the Alhambra at Granada in Spain. The programme is described in the related *Television Notes*.

4   Polyhedra

In Section 1 we discussed some general aspects of geometric structure, and introduced the basic geometric elements in terms of which geometric systems are described — the convex hulls of finite sets of points. In the plane, these are the convex polygons.

Special among the convex polygons are the *regular polygons*, of which there are infinitely many. However, their three-dimensional analogues, the *regular polyhedra*, are only five in number. In this section, we see why this is so, and we also consider their close relatives, the *semi-regular polyhedra*.

4.1 Regular and semi-regular polyhedra

Regular polyhedra

Consider the cube shown in the margin. The faces are all congruent regular polygons (squares), and the same number of them (three) meet at each vertex. These are the defining features of a *regular polyhedron*.

cube

> ### Definition
>
> A **regular polyhedron** is a convex polyhedron in which all the polygonal faces are congruent regular polygons, and in which each vertex has exactly the same arrangement of polygons around it.

According to our definition, all regular polyhedra are convex. Some texts use a slightly different definition that allows certain non-convex objects; we shall not consider these here.

Note that at least three polygons must meet at each vertex, for if only two polygons P and Q were to meet at a vertex v, then the two edges of P incident with v would coincide with the two edges of Q incident with v, and so P and Q would lie in the same plane and could not enclose a volume. This observation will be useful for the next problem.

Problem 4.1

Determine which congruent regular polygons can fit in edge-to-edge contact around a vertex of a regular polyhedron, so that there are no gaps or overlaps.

Hint First explain why the sum of the interior angles of the polygons arranged around a vertex is less than 2π; then use the table in the margin, which reminds you of the interior angles θ of the n-sided regular polygons, for $n = 3, 4, 5, 6$.

n	3	4	5	6
θ	$\pi/3$	$\pi/2$	$3\pi/5$	$2\pi/3$

In Problem 4.1, we determined the five possible **vertex types** that can occur in a regular polyhedron. What regular polyhedra can be formed from them?

From the definition, we require every vertex to have exactly the same arrangement of polygons around it; that is, we require every vertex to have the same vertex type. So there can be at most five regular polyhedra, corresponding to the five vertex types. It is perhaps surprising that all five do exist: they are as follows.

tetrahedron   octahedron   cube   icosahedron   dodecahedron

You have seen these polyhedra before, in *Graphs 1* and in the Introduction to this unit, referred to as the **Platonic solids**, since Plato (*c*.428–*c*.347 BC) made extensive studies of them. They have been of interest to mathematicians, such as Euclid, Kepler and Euler, for thousands of years.

The Platonic solids are discussed in Plato's *Timaeus* and in Book XIII of Euclid's *Elements*.

The numbers of vertices, edges and faces of the Platonic solids are listed below.

polyhedron	vertices	edges	faces
tetrahedron	4	6	4
octahedron	6	12	8
cube	8	12	6
icosahedron	12	30	20
dodecahedron	20	30	12

The ending *hedron* comes from the Greek for 'face'; *tetra* means 4, *octa* means 8, *dodeca* means 12, and *icosa* means 20. The word *cube* comes from *kubos*, the Greek word for die.

There is a duality construction for convex polyhedra, similar to that for tilings. The **dual** of a convex polyhedron is constructed by placing a new vertex at the centre of each face of the original polyhedron, and joining a pair of new vertices with a line segment whenever the corresponding faces of the original polyhedron are edge-adjacent; thus the roles of the vertices and faces are exchanged. For example, the dual of a cube is an octahedron, as shown below.

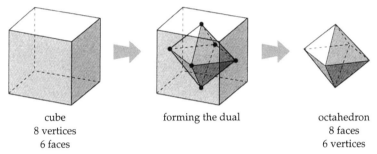

cube
8 vertices
6 faces

forming the dual

octahedron
8 faces
6 vertices

In fact, the duals of all the Platonic solids are themselves Platonic solids.

Problem 4.2

Use the table above to identify the dual of each of the Platonic solids.

Semi-regular polyhedra

The existence of the semi-regular tilings of the plane suggests that a similar class of semi-regular polyhedra may exist. This is indeed the case; as with the semi-regular tilings, we obtain this class by removing the condition that all the polygons should be congruent.

A class of demi-regular polyhedra also exists, but we do not discuss them here.

Definition

A **semi-regular polyhedron** is a convex polyhedron in which all the polygonal faces are regular polygons, not all congruent, such that each vertex has the same arrangement of polygons around it — that is, such that each vertex type is the same.

There are infinitely many semi-regular polyhedra, comprising two infinite sets (the *prisms* and *antiprisms*) and one finite set of thirteen. The thirteen polyhedra in the latter set have been known from the time of the ancient Greeks, and are usually referred to as the **Archimedean solids**, since Archimedes (*c.*287–212 BC) was reputedly the first person to study them.

A **prism** is constructed by taking two parallel congruent regular polygons, aligned one above the other, and joining corresponding vertices with new edges of length equal to that of the sides of the polygons. This results in an 'equatorial strip' of squares, with a polygonal top and base. The prisms in which the defining polygons are equilateral triangles, squares, regular pentagons and regular hexagons are shown below.

In general, the prism whose defining polygon has n edges is called the **n-prism**.

triangular prism	square prism	pentagonal prism	hexagonal prism

An **antiprism** is constructed in a similar manner from two parallel congruent regular polygons, but in this case the polygons are not aligned; rather, the vertices of the upper polygon are placed above the midpoints of the edges of the lower one. The vertices are then joined in such a way as to produce an 'equatorial strip' of equilateral triangles rather than squares. The antiprisms in which the defining polygons are equilateral triangles, squares, regular pentagons and regular hexagons are shown below.

In general, the antiprism whose defining polygon has n edges is called the **n-antiprism.**

triangular antiprism	square antiprism	pentagonal antiprism	hexagonal antiprism

Problem 4.3

(a) Which of the regular polyhedra is a prism?

(b) Which of the regular polyhedra is an antiprism?

The thirteen Archimedean solids are depicted on the next page, together with the five Platonic solids. They are constructed from various combinations of equilateral triangles, squares, regular pentagons, regular hexagons, regular octagons and regular decagons (ten sides). Unlike the semi-regular tilings, they contain the pentagon and decagon, but not the dodecagon (twelve sides).

Several of the Archimedean solids arise in nature in crystal structures.

Ten Archimedean solids involve just two types of polygon, and the remaining three use three types.

The Archimedean solids are given complicated names, which in most cases refer to the number of faces, the types of face present, or else to the method of construction. For example, the Archimedean solid labelled (a) is called a **truncated tetrahedron**, because it can be obtained by truncating a tetrahedron. This process of **truncation** involves slicing off each vertex to create a new polygonal face in its place. The same process gives the **truncated octahedron** (b), the **truncated cube** (c), the **truncated icosahedron** (d) and the **truncated dodecahedron** (e).

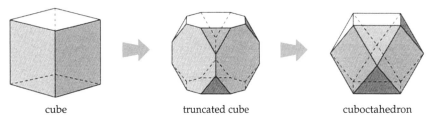

cube	truncated cube	cuboctahedron

If larger slices are used in the truncation, we obtain different solids. Thus the **cuboctahedron** (f) is obtained in this way from either the cube or the octahedron, and is (in a sense) half-way between these two polyhedra (hence the name). Similar remarks apply to the **icosidodecahedron** (g), which lies half-way between an icosahedron and a dodecahedron. The cuboctahedron and icosidodecahedron are characterized by having two kinds of face, with each face of one kind being entirely surrounded by faces of the other kind.

34

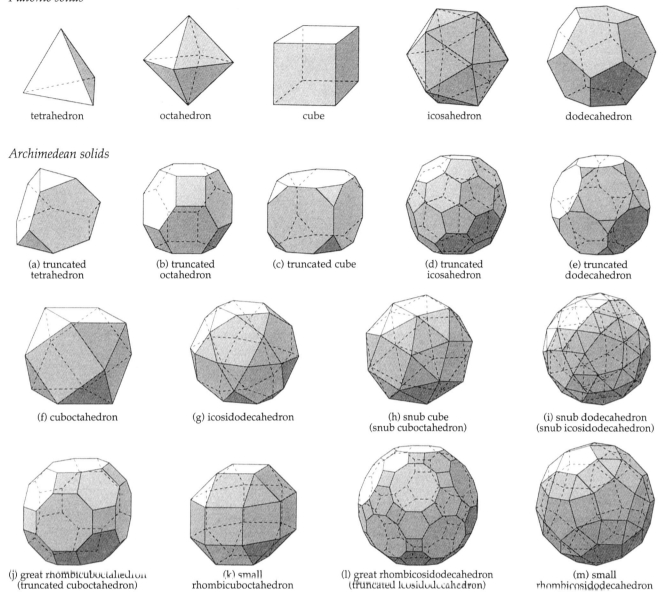

Platonic solids

tetrahedron   octahedron   cube   icosahedron   dodecahedron

Archimedean solids

(a) truncated tetrahedron   (b) truncated octahedron   (c) truncated cube   (d) truncated icosahedron   (e) truncated dodecahedron

(f) cuboctahedron   (g) icosidodecahedron   (h) snub cube (snub cuboctahedron)   (i) snub dodecahedron (snub icosidodecahedron)

(j) great rhombicuboctahedron (truncated cuboctahedron)   (k) small rhombicuboctahedron   (l) great rhombicosidodecahedron (truncated icosidodecahedron)   (m) small rhombicosidodecahedron

The **snub cube** (h), also known as the **snub cuboctahedron**, and the **snub dodecahedron** (i), also known as the **snub icosidodecahedron**, are derived from the cuboctahedron and icosidodecahedron respectively by inserting extra equilateral triangular faces.

Truncation of the cuboctahedron in two different ways gives the **great rhombicuboctahedron** (j), also known as the **truncated cuboctahedron**, and the **small rhombicuboctahedron** (k). Truncation of the icosidodecahedron in two different ways gives the **great rhombicosidodecahedron** (l), also known as the **truncated icosidodecahedron**, and the **small rhombicosidodecahedron** (m).

Problem 4.4

List the numbers of vertices, edges and faces of the hexagonal prism, the truncated tetrahedron and the truncated octahedron.

4.2 Euler's polyhedron formula

Recall from Section 4.1 that the numbers of vertices, edges and faces of the Platonic solids are as follows.

polyhedron	vertices	edges	faces
tetrahedron	4	6	4
octahedron	6	12	8
cube	8	12	6
icosahedron	12	30	20
dodecahedron	20	30	12

There is a simple relationship between the numbers of vertices, edges and faces of any convex polyhedron.

Problem 4.5

For each of the Platonic solids, add the number of vertices and the number of faces, and compare your answer with the number of edges. What do you notice?

The result you discovered for the Platonic solids in Problem 4.5 is true for all convex polyhedra; it is called *Euler's polyhedron formula*, and may be stated formally as follows.

We shall not prove this result here, but will prove a related result in *Graphs 3*.

Theorem 4.1: Euler's polyhedron formula

Let v, e and f denote, respectively, the numbers of vertices, edges and faces of a convex polyhedron. Then

$$v - e + f = 2.$$

Historical note

The polyhedron formula first appeared in this form in a letter from Leonhard Euler to the number theorist Christian Goldbach in November 1750. At that time, Euler was unable to prove the result, but he presented a proof two years later. Unfortunately, Euler's proof was deficient, but a correct proof was obtained by Adrien Marie Legendre in 1794.

It is sometimes claimed that René Descartes obtained the formula around the year 1640; in fact, Descartes obtained an expression for the sum of the angles of all the faces of a polyhedron, from which the result can be deduced, but Descartes apparently never made this deduction.

Problem 4.6

Verify that Euler's polyhedron formula holds for the hexagonal prism, the truncated tetrahedron and the truncated octahedron.

Hint The relevant numbers were obtained in Problem 4.4.

We can use Euler's polyhedron formula to prove results about the existence of regular and semi-regular polyhedra. To do this, we shall also need the following result, which is an analogue of the handshaking lemma for graphs. In order to state it, we define the **degree** of a face of a polyhedron to be the number of edges around it. Thus, a triangular face has degree 3, a square face has degree 4, and so on.

The handshaking lemma for graphs states that, in any graph, the sum of all the vertex degrees is equal to twice the number of edges.

Theorem 4.2: handshaking lemma for polyhedra

In any polyhedron, the sum of all the face degrees is equal to twice the number of edges.

Proof

Since each edge adjoins two faces, it must contribute exactly 2 to the sum of the face degrees. The result follows immediately. ∎

We can now prove the following theorem.

> ## Theorem 4.3
> There are only five regular polyhedra:
> - three with triangular faces — the tetrahedron, the octahedron and the icosahedron;
> - one with square faces — the cube;
> - one with pentagonal faces — the dodecahedron.

We prove the first part of this theorem, and ask you to prove the rest.

Proof

Let v, e and f denote, respectively, the numbers of vertices, edges and faces of a regular polyhedron with triangular faces. It follows from the handshaking lemma for polyhedra that

$$3f = 2e.$$

If exactly d edges meet at each vertex, then it follows from the handshaking lemma for graphs that

$$dv = 2e.$$

These two results can be rewritten as

$$f = 2e/3 \quad \text{and} \quad v = 2e/d.$$

Substituting these two results into Euler's polyhedron formula, $v - e + f = 2$, we obtain

$$2e/d - e + 2e/3 = 2,$$

which (after dividing through by $2e$) can be rewritten as

$$1/d - 1/6 = 1/e.$$

Since $1/e > 0$, it follows that

$$1/d > 1/6,$$

and so $d < 6$. This means that the only possible values of d are 3, 4 and 5.

We consider each case in turn.

Case 1 $d = 3$: then $1/e = 1/3 - 1/6 = 1/6$, so $e = 6$;
  it follows that $f = 4$ and $v = 4$ — this gives the tetrahedron.

Case 2 $d = 4$: then $1/e = 1/4 - 1/6 = 1/12$, so $e = 12$;
  it follows that $f = 8$ and $v = 6$ — this gives the octahedron.

Case 3 $d = 5$: then $1/e = 1/5 - 1/6 = 1/30$, so $e = 30$;
  it follows that $f = 20$ and $v = 12$ — this gives the icosahedron. ∎

Problem 4.7 ────────────────────────────────────

By imitating the above proof, show that:

(a)   the cube is the only regular polyhedron with square faces;

(b)   the dodecahedron is the only regular polyhedron with pentagonal faces.

───

We conclude this section by demonstrating how Euler's polyhedron formula can be used to deduce how many faces of a given degree certain semi-regular polyhedra can possess.

Example 4.1

We shall show that, if a semi-regular polyhedron has only square and hexagonal faces and if exactly three faces meet at each vertex, then it must have exactly six square faces.

Let v, e and f denote, respectively, the numbers of vertices, edges and faces of a semi-regular polyhedron with s square faces and h hexagonal faces. Then

$$f = s + h. \tag{4.1}$$

By the handshaking lemma for polyhedra, the sum of the face degrees (each of which is 4 or 6) is twice the number of edges, and so

$$2e = 4s + 6h; \tag{4.2}$$

that is,

$$e = 2s + 3h. \tag{4.3}$$

Since exactly three faces meet at each vertex, it follows from the handshaking lemma for graphs that

$$3v = 2e.$$

Substituting into equation 4.2, we obtain

$$3v = 4s + 6h,$$

and so

$$v = 4s/3 + 2h. \tag{4.4}$$

Substituting the expressions for v, e and f given by equations 4.4, 4.3 and 4.1 into Euler's polyhedron formula, we obtain

$$(4s/3 + 2h) - (2s + 3h) + (s + h) = 2.$$

The terms involving h cancel, leaving

$$s/3 = 2,$$

and so $s = 6$.

Thus, any polyhedron of this type must have exactly six square faces. ∎

We could continue the analysis in Example 4.1 by substituting $s = 6$ into equations 4.4, 4.3 and 4.1 to obtain

$$v = 8 + 2h, \qquad e = 12 + 3h, \qquad f = 6 + h.$$

Each positive integer value for h then gives us a potential set of numbers of vertices, edges and faces in a semi-regular polyhedron with six square faces, h hexagonal faces and exactly three faces meeting at each vertex. It can be shown, using geometrical considerations beyond the scope of this unit, that only two of these potential sets of numbers give rise to semi-regular polyhedra:

$s = 6$, $h = 2$, $v = 12$, $e = 18$, $f = 8$ gives the hexagonal prism;

Recall your solution to Problem 4.4.

$s = 6$, $h = 8$, $v = 24$, $e = 36$, $f = 14$ gives the truncated octahedron.

Other combinations of polygons and other numbers of faces meeting at a vertex can be analysed in a similar manner.

4.3 Computer activities

The computer activities for this section are described in the *Computer Activities Booklet*.

5 Incidence structures

In the Introduction, we mentioned a design problem involving the testing of drugs on volunteers and another involving the organization of a three-round chess match between two teams of seven players. In this section, we return to these problems, and to graphs and tilings, and put them all in the more general context of an *incidence structure*.

5.1 Examples of incidence structures

The concept of an *incidence structure* is very general, and is useful whenever there are entities of different types that are related to each other.

A familiar example arises in Euclidean geometry. Euclid's axioms for plane geometry concern *points* and *lines* in the plane, and include statements such as:

(a) two distinct points lie on exactly one line;

(b) two distinct non-parallel lines meet in exactly one point.

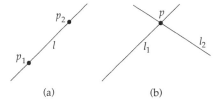

(a) (b)

These essentially concern the properties of an *incidence relation* between the points and the lines: a given point p is incident with a given line l if l passes through p.

Example 5.1: graphs

A graph has vertices and edges, and there is an *incidence relation* between them — that is, for any vertex v and any edge e, we can answer the question 'are v and e incident?' Moreover, the incidence relation tells us everything about the graph. Each edge joins the two vertices with which it is incident, and two vertices are adjacent if and only if there is an edge with which they are both incident. For example, if we know that, in a graph G, with three vertices, vertex v_1 is incident only with edges e_1 and e_2, vertex v_2 is incident only with edge e_1, and vertex v_3 is incident only with edge e_2, then G must be the graph depicted in the margin. ∎

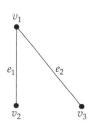

Example 5.2: tilings

In an edge-to-edge tiling with polygons, there are three sets to be considered: vertices, edges and tiles. The vertices and edges that are *incident* with a tile T are those that define the boundary of T, and the vertices at each end of an edge are *incident* with that edge. For example, in the diagram in the margin, the vertices v_1, v_2 and v_3 are incident with the tile T, as are the edges e_1, e_2 and e_3. Also, v_1 is incident with e_1 and e_2, v_2 is incident with e_1 and e_3, and v_3 is incident with e_2 and e_3. Thus, for each pair of sets (tiles and vertices, tiles and edges, vertices and edges), there is an incidence relation. ∎

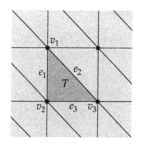

We mentioned in the Introduction that the answer to a drug-testing problem may involve a geometric structure. We now give a specific example.

Example 5.3: drug testing

Suppose that we have seven hay-fever drugs, called *Ampersand, Bastasneeze, Calamitine, Dripnomore, Eezybreathe, Fantastamine* and *Gladeyes*. Seven volunteers present themselves: Alex, Barbara, Chris, Deirdre, Eric, Fred and Gill. The hay-fever season is not long enough for us to ask every tester to try every drug, but it would be useful for each *pair* of drugs to be compared directly by some tester.

This time, the problem is to construct a suitable *incidence relation* between drugs and volunteers that will tell us which drugs to assign to which volunteers. (For instance, if *Ampersand* is incident with Eric in the relation, then one of the drugs Eric will test is *Ampersand*.)

Suppose we decide that:

Alex	should test	*Ampersand, Bastasneeze* and *Dripnomore;*
Barbara	should test	*Bastasneeze, Calamitine* and *Eezybreathe;*
Chris	should test	*Calamitine, Dripnomore* and *Fantastamine;*
Deirdre	should test	*Dripnomore, Eezybreathe* and *Gladeyes;*
Eric	should test	*Eezybreathe, Fantastamine* and *Ampersand;*
Fred	should test	*Fantastamine, Gladeyes* and *Bastasneeze;*
Gill	should test	*Gladeyes, Ampersand* and *Calamitine.*

Then, each pair of drugs is directly compared by just one of the testers in this scheme. ∎

Problem 5.1

Verify that *Ampersand* is directly compared with each other drug by exactly one tester.

Each of these examples gives an *incidence structure*, which we define formally as follows.

Definition

An **incidence structure** $S(P, L)$ consists of two sets, P and L, of different types of object, and an **incidence relation** that tells us, for each pair of objects, one from P and one from L, whether or not these two objects are incident.

We often describe the elements of the two sets P and L as 'points' and 'lines' respectively, since a geometry of points and lines is an archetypal incidence structure. We denote the numbers of 'points' and 'lines' by n_p and n_l respectively. Alternatively, for examples related to a graph G, we may label the sets V and E, refer to them as 'vertices' and 'edges', and write $G(V, E)$.

We sometimes use quotation marks for the 'points' and 'lines' of an incidence structure, to emphasize that they are not necessarily the familiar points and lines of a Euclidean space — the names are just there to draw our attention to the geometric aspects of the structure.

Definitions

The **degree** of a 'point' is the number of 'lines' incident with it, and the **degree** of a 'line' is the number of 'points' incident with it.

An incidence structure is **regular** if every 'point' has the same degree d_p and every 'line' has the same degree d_l.

This definition of *degree* generalizes the definition of the degree of a vertex of a simple graph. (The degree of any edge of a simple graph is 2, since each edge has two vertices incident with it.)

Our faithful cuboid provides six examples of regular incidence structures, three of which are the following:

* taking the vertices as the 'points' and the edges as the 'lines', we have a regular incidence structure S_1 with $n_p = 8$, $n_l = 12$, $d_p = 3$ and $d_l = 2$.

- taking the faces as the 'points' and the edges as the 'lines', we have a regular incidence structure S_2 with $n_p = 6$, $n_l = 12$, $d_p = 4$ and $d_l = 2$.

- taking the vertices as the 'points' and the faces as the 'lines', we have a regular incidence structure S_3 with $n_p = 8$, $n_l = 6$, $d_p = 3$ and $d_l = 4$.

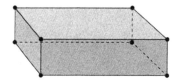

8 vertices of degree 3
12 edges of degree 2
6 faces of degree 4

Problem 5.2

What are the other three examples of regular incidence structures provided by the cuboid?

Problem 5.3

Verify that, for each of the six cuboid examples, $n_p d_p = n_l d_l$.

We conclude this subsection by introducing a concept of duality that is more abstract and more general than the concepts of *dual tiling* and *dual polyhedron* that you met in Sections 2 and 4.

> **Definition**
>
> Let $S(P, L)$ be an incidence structure. Then the incidence structure $S^*(L, P)$ obtained by exchanging the roles of P and L is called the **dual** of $S(P, L)$.

For example, the incidence structure S_4 obtained by taking the edges of the cuboid as 'points' and the vertices as 'lines' is the dual of the incidence structure S_1 obtained by taking the vertices as 'points' and the edges as 'lines'. Similarly S_1 is the dual of S_4.

Problem 5.4

What are the duals of the other four examples of regular incidence structures provided by the cuboid?

Hint   See the solution to Problem 5.3.

5.2   Representing incidence structures

In *Graphs 1*, we showed how a graph can be represented by an incidence matrix whose rows correspond to the vertices, and whose columns correspond to the edges. In such a matrix, a 1 appears whenever the corresponding vertex and edge are incident, and a 0 appears otherwise.

Example 5.1: graphs

In this case, we obtain the following incidence matrix:

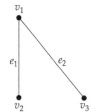

$$\begin{array}{c} & \begin{array}{cc} e_1 & e_2 \end{array} \\ \begin{array}{c} v_1 \\ v_2 \\ v_3 \end{array} & \left[\begin{array}{cc} 1 & 1 \\ 1 & 0 \\ 0 & 1 \end{array} \right] \end{array}$$

Recall that the sum of the 1s in each row is the degree of the corresponding vertex and that the sum of the 1s in each column is 2 — the degree of the corresponding edge.

■

Similarly, we can represent other incidence structures by means of an incidence matrix.

Example 5.2: tilings

In this case, an incidence matrix pertinent to tile T, in which the entry in row i and column j is 1 if vertex v_i is incident with edge e_j, and 0 otherwise, is as follows:

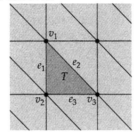

$$
\begin{array}{c}
 & \begin{array}{ccc} e_1 & e_2 & e_3 \end{array} \\
\begin{array}{c} v_1 \\ v_2 \\ v_3 \end{array} &
\begin{bmatrix}
1 & 1 & 0 \\
1 & 0 & 1 \\
0 & 1 & 1
\end{bmatrix}
\end{array}
$$

∎

Example 5.3: drug testing

In this example, the information about who tests which drug can be represented by the following incidence matrix, in which the drugs are represented by their initials and the entry in row i and column j is 1 if volunteer i tests drug j, and 0 otherwise:

	A	B	C	D	E	F	G
Alex	1	1	0	1	0	0	0
Barbara	0	1	1	0	1	0	0
Chris	0	0	1	1	0	1	0
Deirdre	0	0	0	1	1	0	1
Eric	1	0	0	0	1	1	0
Fred	0	1	0	0	0	1	1
Gill	1	0	1	0	0	0	1

∎

The sum of the 1s in each row is the number of drugs tested by the corresponding volunteer and the sum of the 1s in each column is the number of times the corresponding drug is tested.

Each of these examples gives an *incidence matrix*, which we define formally as follows.

> ## Definition
>
> Let $S(P, L)$ be an incidence structure, with n_p 'points' labelled 1 to n_p and n_l 'lines' labelled 1 to n_l. The **incidence matrix** $\mathbf{B}(S)$ is the $n_p \times n_l$ matrix in which the entry in row i and column j is 1 if the 'point' labelled i is incident with the 'line' labelled j, and 0 otherwise.

Problem 5.5

The following figures represent a cuboid and its graph: the vertices are labelled 1, 2, ... , 8; the front face is labelled 1, the back face 2, the left face 3, the right face 4, the top face 5 and the bottom face 6.

In the graph, the back face 2 has become an 'infinite face'.

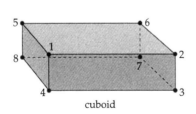

cuboid

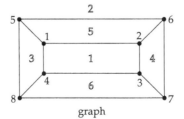

graph

(a) Write down the incidence matrix of the incidence structure whose 'points' are the vertices and whose 'lines' are the faces.

(b) Write down the incidence matrix of the dual incidence structure.

(c) What connection do you notice between the incidence matrices obtained in parts (a) and (b)?

In the solution to Problem 5.5, we introduced the concept of the *transpose* of a matrix. In case you have not met this idea before, we now define it

Definition

Let **A** be any matrix with m rows and n columns. Then the **transpose** of **A**, denoted by $\mathbf{A}^T$, is the matrix with n rows and m columns obtained by interchanging the rows and columns of **A**; that is, the entry in row i and column j of **A** is the same as the entry in row j and column i of $\mathbf{A}^T$.

For example, the transpose of

$$\begin{bmatrix} 1 & 1 \\ 1 & 0 \\ 0 & 1 \end{bmatrix} \text{ is } \begin{bmatrix} 1 & 1 & 0 \\ 1 & 0 & 1 \end{bmatrix}.$$

We now generalize the results of Problems 5.5(c) and 5.3.

Theorem 5.1

Let S be any incidence structure, and let S^* be its dual. Then:

(a) the incidence matrix of S^* is the transpose of that of S;

(b) the sum of the degrees of the 'points' of S is equal to the sum of the degrees of its 'lines'.

Part (b) is a generalization of the handshaking lemma for graphs, of the handshaking lemma for polyhedra and of the result of Problem 5.3.

Proof

(a) The 'point' i and 'line' j of S are the 'line' i and 'point' j of S^*; thus the entry in row i and column j of the incidence matrix of S is equal to the entry in row j and column i of the incidence matrix of S^*.

(b) Each of these numbers is equal to the total number of point–line incidences — that is, to the total number of 1s in the incidence matrix. ∎

Problem 5.6 ⎯⎯⎯⎯⎯⎯⎯⎯⎯⎯⎯⎯⎯⎯⎯⎯⎯⎯⎯⎯⎯⎯⎯

(a) Prove that the equation $n_p d_p = n_l d_l$ holds for any regular incidence structure.

(b) Deduce the handshaking lemma for graphs and for polyhedra from Theorem 5.1(b).

Another useful way of representing an incidence structure is by means of a **block table**. This consists of a set of columns, each of which represents a 'line', by containing a list those 'points' with which it is incident.

This representation will be used extensively in *Design 4*.

In the drug-testing incidence structure, for example, the first 'line' Alex is incident with the 'points' *Ampersand*, *Bastasneeze* and *Dripnomore*, and so the first column of the table contains the letters A, B and D. The full block table is:

A	B	C	D	E	F	G
A	*B*	*C*	*D*	*E*	*F*	*G*
B	*C*	*D*	*E*	*F*	*G*	*A*
D	*E*	*F*	*G*	*A*	*B*	*C*

We use open-face letters to represent the volunteers ('lines'). Within each column, the order of the italic letters representing the drugs ('points') is unimportant.

Problem 5.7 ⎯⎯⎯⎯⎯⎯⎯⎯⎯⎯⎯⎯⎯⎯⎯⎯⎯⎯⎯⎯⎯⎯⎯

(a) Write down the block table for the drug-testing example, taking the testers as the 'points' and the drugs as the 'lines'.

(b) In the vertex–face incidence structure in Problem 5.5(a), the first 'line' 1 is incident with the four 'points' 1, 2, 3, 4, and so the first column of the block table contains the numbers 1, 2, 3 and 4. Write down the complete block table.

5.3 Finite projective geometry

Geometries obeying axioms other than those of Euclid were discovered in the nineteenth century. It is not the place of this unit to go into the fascinating history of non-Euclidean geometry and its effect on the philosophy of mathematics. However, it is worth recording that the crucial difference between Euclidean and non-Euclidean geometry lies in the properties of the incidence relation between 'points' and 'lines'.

The original non-Euclidean geometries were conceived of as being somewhat like Euclid's geometry except for differences involving parallel lines. In particular, like the Euclidean plane, they contained infinitely many 'points' and 'lines'. Later, however, it was noticed that *finite* sets of 'points' and 'lines' can be devised that satisfy the various axioms of non-Euclidean geometry. These sets are called **finite geometries**.

Example 5.4: the Fano plane

The **Fano plane** has seven points and seven lines, each line being incident with three points, and each point with three lines. It can be drawn as follows:

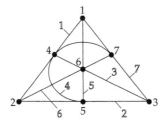

Six of the lines are drawn as line segments, while one is drawn as a curve.

The block table of the Fano plane can be written as follows:

1	2	3	4	5	6	7
1	2	3	4	5	6	7
2	3	4	5	6	7	1
4	5	6	7	1	2	3

Notice that we have written the elements of the columns in a particular order to show the structure of the table: each column is obtained from the previous one by adding 1 to each number (replacing 8 by 1). ∎

This 'cyclic' property of certain block tables is discussed in *Design 4*.

Problem 5.8

The Fano plane has been described by the nonsense sentence

YEA, WHY TRY HER RAW WET HAT.

By assigning points to the seven letters, show that each line corresponds to a word of this sentence.

There are several types of finite geometry. The Fano plane is an example of a type known as a *finite projective plane*.

We shall not explain how the phrase 'projective plane' arises, as this would take us too far afield.

Definition

A **finite projective plane** is a regular incidence structure in which:

(a) for each pair of distinct 'points', there is exactly one 'line' incident with both;

(b) for each pair of distinct 'lines', there is exactly one 'point' incident with both.

Further restrictions are usually imposed in order to avoid trivial structures, such as that consisting of just one 'line' incident with an arbitrary number of 'points'. For our purposes, however, this definition suffices.

Problem 5.9

Verify that the Fano plane is a finite projective plane.

It has been shown that any finite projective plane must have equal numbers of 'points' and 'lines', and that these numbers must be of the form $k^2 + k + 1$, for some integer k. Moreover, there are exactly $k + 1$ 'points' incident with each 'line' and exactly $k + 1$ 'lines' incident with each 'point'. The Fano plane corresponds to the case $k = 2$. The following diagram shows a finite projective plane with 13 points and 13 lines (corresponding to $k = 3$).

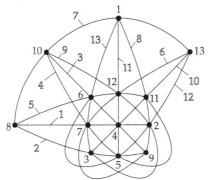

Finite projective planes exist when $k = 2, 3, 4, 5, 7, 8$ and 9 but not when $k = 6$. The case $k = 10$ remained unsolved until 1988, when a long computer search established that no such structure can exist.

Note that the incidence structure of the drug-testing scheme of Example 5.3 is the Fano plane in disguise! The easiest way to see this is to look at the block table:

A	B	C	D	E	F	G
A	B	C	D	E	F	G
B	C	D	E	F	G	A
D	E	F	G	A	B	C

Substituting 1 for A, 2 for B, and so on, we obtain the block table of the Fano plane, given in Example 5.4. The advantage of such an arrangement is that each pair of drugs is compared by one of the volunteers, and each pair of volunteers shares one of the drugs. Thus any differences between the drugs have a good chance of being identified, and any volunteer with unusual characteristics is likely to be easily identified as well.

Finally, we return to another question posed in the Introduction.

Problem 5.10

Show how the Fano plane can be used for a three-round chess tournament between two teams, A and B, of seven players each.

After studying this section, you should be able to:

- explain the terms *incidence structure*, *regular incidence structure* and *finite projective plane*, and give examples of these;
- construct the *dual* of an incidence structure;
- represent any of these structures by means of an *incidence matrix* or a *block table*;
- state and prove a theorem on incidence structures that in part generalizes the handshaking lemma.

Further reading

An excellent book that covers much of the material in this unit is:

P. Pearce, *Structure in Nature is a Strategy for Design*, MIT Press, 1978.

A detailed account of geometric elements is given in:

H. S. M. Coxeter, *Regular Polytopes*, Dover Publications, 1973.

The definitive work on tilings and related subjects is:

B. Grünbaum and G. C. Shephard, *Tilings and Patterns*, Freeman Publications, 1988.

A catalogue of regular, semi-regular and demi-regular tilings, each presented on a separate A4 page, is given in:

S. Bezuska, M. Kenney and L. Silvey, *Tessellations, the Geometry of Patterns*, Creative Publications, 1977.

One of the best sources of information on the polygonal animals is:

S. W. Golomb, *Polyominoes*, Charles Scribner, 1965.

An interesting source book for polyhedra, which describes various methods of construction, is:

H. M. Cundy and A. P. Rollett, *Mathematical Models*, Oxford University Press, 1961.

A more recent book on the same subject, which has a bias towards applications in architecture, is:

A. Pugh, *Polyhedra, a Visual Approach*, University of California at Berkeley Press, 1976.

Acknowledgements

p.15 photographs of Liverpool Roman Catholic Cathedral, the London Ark, the Atomium in Brussels, and the Pyramid at the Louvre in Paris, courtesy of the Architectural Association;

p.22 photograph of bees' honeycomb, courtesy of Heather Angel Biofotos;

p.23 photograph of the Royal Bathroom in the Alhambra, courtesy of J. Allan Cash Limited;

p.23 close-up photographs of two tilings in the Alhambra, courtesy of Ampliaciones Reproducciones MAS.

Exercises

Section 1

1.1

(a) State the dimension of the circle of points in the plane that satisfy the equation

$$x^2 + y^2 = 1.$$

(b) Suggest a suitable way of describing the points belonging to the circle, using the appropriate number of coordinates.

1.2 Demonstrate that each of the following types of plane segment can be obtained by overlapping two triangles:

(a) a triangle;

(b) a square;

(c) a pentagon.

1.3 In the plane, let $P = (0, 0)$, $Q = (2, 0)$, $R = (0, 2)$, $S = (1, 1)$, $T = (2, 2)$. Describe the convex hull of each of the following sets of points:

(a) $\{P, Q, R, S, T\}$; (b) $\{P, Q, R, S\}$; (c) $\{Q, R, S\}$; (d) $\{Q, S, T\}$.

1.4 Which of the following shapes are convex and which are non-convex? Which of the convex shapes is the convex hull of a finite set of points? For each non-convex shape, find a line segment that demonstrates non-convexity.

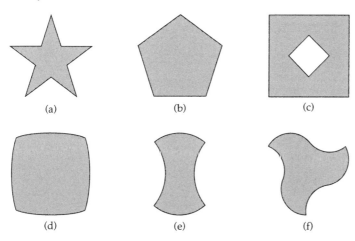

1.5 Let S and T be two overlapping convex sets. Show that their overlap, $S \cap T$ (the set of points common to S and T), is also a convex set.

Section 2

2.1 Which of the following tilings are edge-to-edge?

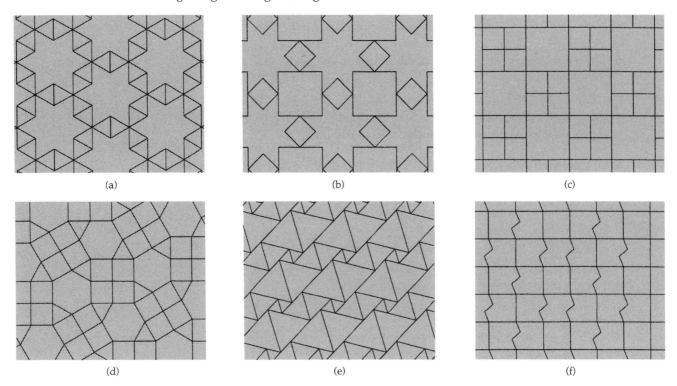

(a) (b) (c)

(d) (e) (f)

2.2 Let *ABC* be any triangle. Show that a triangle *BCD* can be found, congruent to *ABC*, so that *ABDC* is a parallelogram. Hence show that *any* triangle tiles the plane.

2.3 List the codes for each of the following demi-regular tilings:

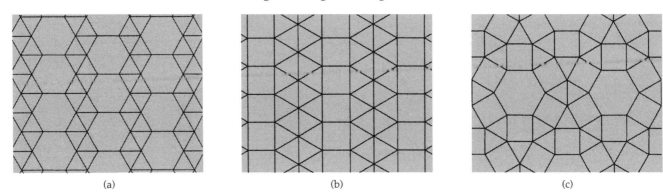

(a) (b) (c)

2.4 Explain why there is no semi-regular tiling with vertex type 3.4.4.6.

Section 3

There are no exercises for Section 3.

Section 4

4.1 Let *R* be a regular polyhedron with v vertices, each of degree d_v, and f faces, each of degree d_f. It is truncated so as to produce a semi-regular polyhedron *S*, each of whose vertices is of degree 3. Determine the number of faces of *S*, and the degree of each face.

4.2 Let *R* be a regular polyhedron with v vertices, each of degree d_v, and f faces, each of degree d_f. Use Euler's polyhedron formula to show that

$$1/d_v + 1/d_f > 1/2.$$

4.3 Using a code analogous to that used for the plane tilings, describe the arrangement of polygons around the vertices of each of the regular polyhedra and around the vertices of the truncated tetrahedron, cuboctahedron and small rhombicuboctahedron.

4.4 Using a code analogous to that used for the plane tilings, describe the arrangement of polygons around the vertices of the n-prism and the n-antiprism.

Section 5

5.1 A food company wishes to test six varieties of sausage, labelled A, B, C, D, E, F. Eight volunteers, labelled $1, 2, 3, 4, 5, 6, 7, 8$, are each willing to sample three varieties.

(a) Find an allocation of sausages to volunteers, based on a Platonic solid.

(b) Is each pair of varieties compared directly by some volunteer? If not, how many pairs are not so compared?

Hint Which Platonic solid has six faces and eight vertices?

5.2 Let G be the following graph of a cube:

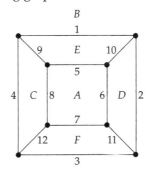

The *edges* are labelled 1 to 12, and the faces are labelled A to F. (The infinite face is labelled B.)

Each adjacent pair of faces defines a 6-cycle, whose edges are the edges adjacent to *just one* of the pair. (For example, the pair of faces (A, D) defines the 6-cycle 5, 10, 2, 11, 7, 8; the edge 6 is not part of the 6-cycle as it is adjacent to both A and D.)

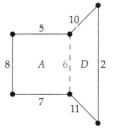

Let S be the incidence structure whose 'points' are these 6-cycles and whose 'lines' are the edges of G, a 'point' being incident with a 'line' if the corresponding 6-cycle contains the corresponding edge.

(a) Find n_p, n_l, d_p and d_l.

(b) The adjacent pairs of faces can be numbered 1 to 12 by the edges that separate them. (For example, the pair of faces (A, D) is separated by edge 6, and so can be numbered 6.) Using this numbering, construct the block table for S. (For example, column 6 contains the numbers 5, 10, 2, 11, 7, 8.)

5.3 A health food company wishes to test n varieties of nut roast mixture, and has a panel of volunteers each prepared to sample k varieties (where $k < n$). If each pair of varieties is to be compared directly by some volunteer, show that at least $n(n-1)/k(k-1)$ volunteers are needed.

5.4 Write down the block table of the finite projective plane with 13 points and 13 lines.

Solutions to the exercises

1.1

(a) The dimension is 1, because y can be expressed in terms of x.

(b) These points all have polar coordinates $(1, \theta)$, where $0 \leq \theta < 2\pi$, so they can be described by the single coordinate θ.

1.2

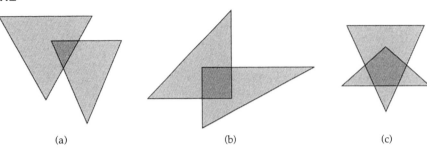

(a)   (b)   (c)

1.3

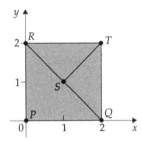

(a) $\langle P, Q, R, S, T \rangle$ is the square $PQTR$;

(b) $\langle P, Q, R, S \rangle$ is the triangle PQR;

(c) $\langle Q, R, S \rangle$ is the line segment QR;

(d) $\langle Q, S, T \rangle$ is the triangle QST.

1.4 Only shapes (b) and (d) are convex.

The only convex shape that is the convex hull of a finite set of points is (b), which is the convex hull of its five corners.

The non-convex shapes (a), (c), (e) and (f) have (for example) the following line segments that demonstrate non-convexity:

In shape (d), no point on the boundary lies on a line segment joining any other pair of points. Therefore, all the infinitely many points on the boundary are required in order to define the shape as a convex hull.

 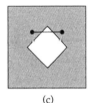  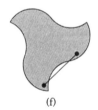

(a)   (c)   (e)   (f)

1.5 Let P and Q be any two points in $S \cap T$. Since S is convex, the line segment PQ lies wholly in S. Similarly, the line segment PQ lies wholly in T. Therefore it lies wholly in $S \cap T$, and so $S \cap T$ fulfils the condition for convexity.

2.1 Tilings (a) and (d) are edge-to-edge; the others are not.

2.2 If triangle BCD is obtained from triangle ABC by rotating through π about the midpoint of BC, then the angles α shown in the figure below are equal, as are the angles β. Also, AB is equal to CD in length, and AC is equal to BD. Thus $ABDC$ is a parallelogram.

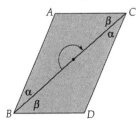

Now, congruent parallelograms can be laid side by side to form an infinite strip, and infinite strips can be laid alongside each other so as to tile the plane. Therefore, since we started with any triangle ABC, it follows that any triangle tiles the plane.

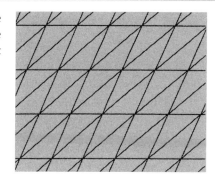

2.3 The codes are:

(a) 3.3.3.3.6/3.3.6.6.

(b) 3.3.3.3.3.3/3.3.3.4.4.

(c) 3.3.4.3.4/3.4.6.4.

2.4 Consider any equilateral triangle ABC. If this were part of a semi-regular tiling with vertex type 3.4.4.6 then we must have the following arrangement of tiles at A (or its reflection):

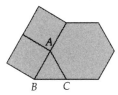

Then, at B, we must have:

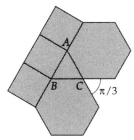

The remaining angle at C is $\pi/3$, and so the only possible vertex type at C is 3.6.3.6. Hence a semi-regular tiling with vertex type 3.4.4.6 is impossible.

4.1 If we truncate a vertex of degree d_v, we replace it by a face of degree d_v. If the new vertices are of degree 3, then each is incident with part of an old edge and with two new edges. Each old face is converted into a face of *double* the degree.

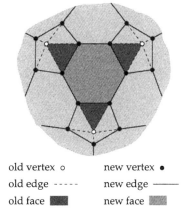

old vertex ○   new vertex ●
old edge - - - -   new edge ———
old face ▪   new face ▨

Thus, S has $v + f$ faces, of which v are of degree d_v and f are of degree $2d_f$.

4.2 By the handshaking lemma for polyhedra, we have

$$fd_f = 2e.$$

It follows from the handshaking lemma for graphs that

$$vd_v = 2e.$$

The two results can be rewritten as

$$v = 2e/d_v \quad \text{and} \quad f = 2e/d_f.$$

Substituting into Euler's polyhedron formula, we get

$$2e/d_v - e + 2e/d_f = 2,$$

which (after dividing through by $2e$) can be rewritten as

$$1/d_v + 1/d_f = 1/e + 1/2.$$

Since $1/e > 0$, it follows that

$$1/d_v + 1/d_f > 1/2.$$

4.3

tetrahedron: 3.3.3

octahedron: 3.3.3.3

cube: 4.4.4

icosahedron: 3.3.3.3.3

dodecahedron: 5.5.5

truncated tetrahedron: 3.6.6

cuboctahedron: 3.4.3.4

small rhombicuboctahedron: 3.4.4.4

4.4

n-prism: 4.4.n

n-antiprism: 3.3.3.n

5.1

(a) The cube is a Platonic solid with six faces and eight vertices. Let us label them as follows:

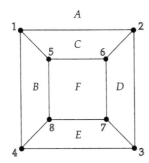

We then use the incidences in the above graph to allocate sausages to volunteers as follows:

1	2	3	4	5	6	7	8
A	A	A	A	B	C	D	B
B	C	D	B	C	D	E	E
C	D	E	E	F	F	F	F

(b) No. Only pairs of faces that share a vertex correspond to pairs of varieties that are directly compared. The three pairs (A, F), (B, D), (C, E) are not so compared.

5.2

(a) There are twelve adjacent pairs of faces, and twelve edges, so

$$n_p = n_l = 12.$$

Each 'point' is incident with six 'lines', as each 6-cycle involves six edges, and so

$$d_p = 6.$$

Therefore, since $n_p d_p = n_l d_l$, we also know that

$$d_l = 6.$$

(b)

1	2	3	4	5	6	7	8	9	10	11	12
2	1	1	1	1	2	3	4	1	1	2	3
3	3	2	2	6	5	5	5	4	2	3	4
4	4	4	3	7	7	6	6	5	5	6	7
5	6	7	8	8	8	8	7	8	6	7	8
9	10	11	9	9	10	11	9	10	9	10	9
10	11	12	12	10	11	12	12	12	11	12	11

5.3 There are $n(n - 1)/2$ pairs of varieties to be compared. Each volunteer can directly compare any two of the k varieties that he or she samples — that is, $k(k - 1)/2$ pairs. Therefore, the number of volunteers needed is at least

$$\frac{n(n - 1)/2}{k(k - 1)/2} = \frac{n(n - 1)}{k(k - 1)}.$$

5.4

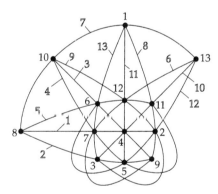

The block table is:

1	2	3	4	5	6	7	8	9	10	11	12	13
2	3	4	5	6	7	8	9	10	11	12	13	1
4	5	6	7	8	9	10	11	12	13	1	2	3
7	8	9	10	11	12	13	1	2	3	4	5	6
8	9	10	11	12	13	1	2	3	4	5	6	7

The order of the numbers in each column does not matter. We have ordered them to show the 'cyclic' pattern: each column is obtained from the previous one by adding 1 to each number (replacing 14 by 1).

Solutions to the problems

Solution 1.1

(a) The line L is one-dimensional, as should be clear intuitively. Mathematically, we can argue as follows. The line L is the set of points whose x- and y-coordinates satisfy the equation $y = x - 1$; that is, the set of points with coordinates $(x, x - 1)$, as x varies. Thus, once the x-coordinate of a point on L is specified, we also know the y-coordinate, so only one of these coordinates is needed in order to describe the position of the point.

(b) The space consisting of the single point P is zero-dimensional.

Solution 1.2

There are numerous possibilities — for example, a brick wall is a man-made system that consists of a large number of repeated units of a single component (the brick). Similarly, a knitted sweater is a man-made system in which the repeated units are the stitches. An example of a natural system is a bees' honeycomb, consisting of essentially identical components (the individual cells) organized in a regular pattern.

Solution 1.3

The vertices are zero-dimensional.

The edges are one-dimensional — each is a subset of a one-dimensional line, so at most one coordinate is needed; on the other hand, the positions of the points on an edge vary along that edge, so one coordinate is indeed needed.

Reasoning in a similar way, we deduce that the faces are two-dimensional and that the cuboid is three-dimensional.

Solution 1.4

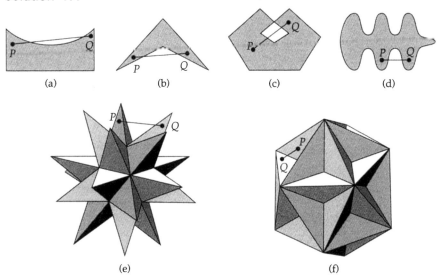

Solution 1.5

If $S = (\frac{1}{3}, \frac{1}{3})$, then S lies inside the triangle PQR, and so $\langle P, Q, R, S \rangle$ is just the triangle PQR.

If $S = (\frac{2}{3}, \frac{2}{3})$, then S lies outside the triangle PQR, and so $\langle P, Q, R, S \rangle$ is the quadrilateral $PQSR$.

If $S = (\frac{1}{2}, \frac{1}{2})$, then S lies on the line segment QR, and so $\langle P, Q, R, S \rangle$ is again the triangle PQR.

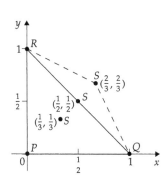

54

Solution 1.6

(a) An example of a quadrilateral is $\langle P, Q, S, T \rangle$. Other quadrilaterals are

$$\langle P, Q, S, U \rangle, \quad \langle P, Q, T, U \rangle, \quad \langle P, S, T, U \rangle, \quad \langle Q, S, T, U \rangle.$$

An example of a non-quadrilateral is the triangle $\langle R, S, T, U \rangle$ — the point U lies inside the triangle RST, and so $\langle R, S, T, U \rangle = \langle R, S, T \rangle$. Other non-quadrilaterals are

$$\langle P, Q, R, S \rangle, \quad \langle P, Q, R, T \rangle, \quad \langle P, Q, R, U \rangle,$$
$$\langle P, R, S, T \rangle, \quad \langle P, R, S, U \rangle, \quad \langle P, R, T, U \rangle,$$
$$\langle Q, R, S, T \rangle, \quad \langle Q, R, S, U \rangle, \quad \langle Q, R, T, U \rangle.$$

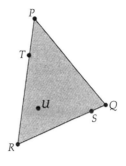

(b) The convex hull of $\langle P, Q, S, T, U \rangle$ is the pentagon $PQSUT$.

Solution 1.7

(a) $\langle P, Q, R, T \rangle$ is the triangle PQR;

(b) $\langle P, T \rangle$ is the line segment PT;

(c) $\langle P, Q, U \rangle$ is the triangle PQU;

(d) $\langle P, Q, R, U \rangle$ is the tetrahedron $PQRU$;

(e) $\langle P, Q, R, T, U \rangle$ is the tetrahedron $PQRU$;

(f) $\langle P, Q, S, T \rangle$ is the triangle PQS.

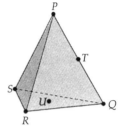

Solution 1.8

An infinite straight line (that is, a one-dimensional Euclidean space) is a convex one-dimensional set that is not a simplex. Another example is a 'half-line', such as the positive half of the x-axis. Similarly, two-dimensional and three-dimensional Euclidean spaces are convex sets that are not simplices, as are 'half-planes' and quadrilaterals in two dimensions, cuboids in three dimensions, and so on.

Solution 1.9

The graphs of the 0-, 1-, 2- and 3-simplices are the complete graphs K_1, K_2, K_3 and K_4, respectively:

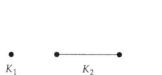

The graph of an n-simplex is the complete graph K_{n+1}.

Solution 1.10

For $n = 1, 2, 3$, the n-orthotopes have 2, 4 and 8 vertices respectively; this suggests that an n-orthotope has 2^n vertices. This is indeed the case: we form an n-orthotope by taking an $(n-1)$-dimensional orthotope (with 2^{n-1} vertices) and translating it to a new position in a direction at right angles to all its $n-1$ dimensional directions — this doubles the number of vertices to $2 \times 2^{n-1} = 2^n$.

Solution 1.11

Each vertex of an n-orthotope (and its graph) has degree n, since the process of translating in a new direction always increases the degree by 1. Thus, by the handshaking lemma, the number of edges is half the sum of the vertex degrees, which is $\frac{1}{2} \times n \times 2^n = n \times 2^{n-1}$.

Solution 1.12

The number of vertices of an n-octahedron is $2n$.

Each vertex is joined to all but one of the others, so the degree of each vertex is $2(n-1)$. Thus, by the handshaking lemma, the number of edges is half the sum of the vertex degrees, which is $\frac{1}{2} \times 2n \times 2(n-1) = 2n(n-1)$.

Solution 2.1

In each case, we superimpose the point lattice, and the corresponding pattern obtained by joining the dots on adjacent bricks, on the packing of bricks.

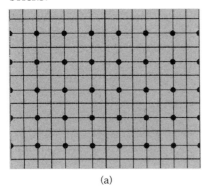

(a)

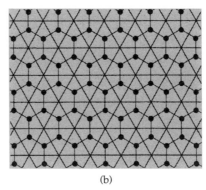

(b)

Solution 2.2

Tilings (b) and (c) are edge-to-edge tilings; (a) and (d) are not.

In tiling (a), each arrow-shaped polygon has two long edges, but these edges are bisected externally, so half of each coincides with an edge of a cross-shaped polygon. Also, each cross-shaped polygon shares *three* edges with each neighbouring arrow-shaped polygon.

In tiling (d), each non-convex polygon shares two edges with each neighbouring convex polygon.

Solution 2.3

(a) Whenever three regular pentagons in edge-to-edge contact meet at a common vertex, then there is an angle of $2\pi - 3(3\pi/5) = \pi/5$ left over, and no other regular pentagon can fill the left-over space.

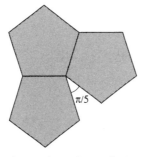

(b) The tiles can be fitted together along their vertical edges to form rows, which can be arranged with the apexes pointing alternately up and down:

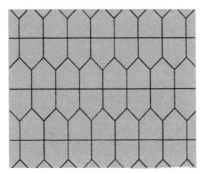

Solution 2.4

If we join the centre of the regular polygon to each of the n vertices with a line segment, then we split the polygon into n triangles.

The sum of all the angles of these triangles is $n\pi$. However, the sum of all the angles around the centre point of the polygon is 2π. Therefore, the sum of all the interior angles around the edges of the polygon is $(n-2)\pi$. Since there are n equal interior angles, each interior angle θ of an n-sided regular polygon is

$$\theta = (n-2)\pi/n.$$

interior angle of a regular hexagon is
$(6-2)\pi/6 = \frac{2}{3}\pi$

Solution 2.5

(b)   3.3.3.4.4/3.3.4.3.4/4.4.4.4;

(c)   3.4.4.6/3.4.6.4/3.6.3.6;   (d)   3.4.4.6/3.4.6.4;

(e)   3.3.3.4.4/3.3.4.3.4;   (f)   3.4.4.6/3.6.3.6.

Solution 2.6

Each of the regular tilings has a regular tiling for a dual. The tilings 3.3.3.3.3.3 and 6.6.6 are duals of each other, while 4.4.4.4 is self-dual. The duals of the semi-regular tilings use tiles that are not regular polygons, and so cannot be regular or semi-regular tilings.

Solution 2.7

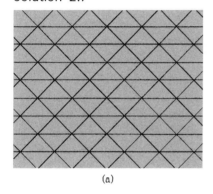

(a)

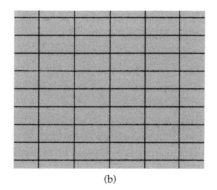

(b)

Solution 2.8

$n = 1$	$n = 2$	$n = 3$	$n = 4$

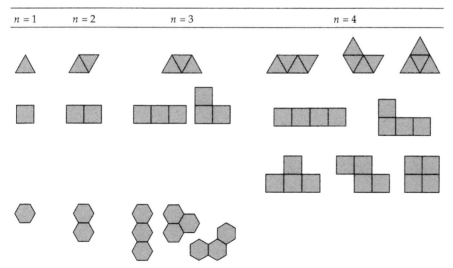

Solution 2.9

The non-isomorphic graphs in each case are as follows:

n-iamonds

n-ominoes

n-hexes

Solution 4.1

The polygons around a vertex of a polyhedron do not lie in the same plane. The sum of the interior angles of the polygons around a vertex cannot therefore be 2π. If this sum were greater than 2π, the vertex would have to be part of a concavity, or indentation, of the polyhedron — but this is excluded since a regular polyhedron is convex. We conclude that the sum of the angles is less than 2π.

The polygons around a vertex of a regular polyhedron are all of the same type, and so have the same interior angles. We therefore seek interior angles such that multiples of three or more of them come to less than 2π. From the table, we see that the only possibilities involve the equilateral triangle, the square and the regular pentagon ($n = 3$, 4 and 5). The regular hexagon ($n = 6$) is excluded, because three hexagons around a vertex lie in a plane. For $n > 6$ the situation is even worse, since three such polygons around a vertex would either overlap or produce a concavity, giving a non-convex polyhedron.

Thus the following *vertex types* can occur.

These diagrams depict non-planar figures: each vertex is peaked out of the plane of the page.

Notice that arrangements of six equilateral triangles or four squares around a vertex are excluded because they lie in a plane.

Solution 4.2

In each case, the dual is the solid with the same number of edges as the original, and with the numbers of vertices and faces exchanged:

original	dual
tetrahedron	tetrahedron
octahedron	cube
cube	octahedron
icosahedron	dodecahedron
dodecahedron	icosahedron

Solution 4.3

(a) The cube is the square prism.

(b) The octahedron is the triangular antiprism.

Solution 4.4

polyhedron	vertices	edges	faces
hexagonal prism	12	18	8
truncated tetrahedron	12	18	8
truncated octahedron	24	36	14

Solution 4.5

In each case, the number of vertices plus the number of faces is equal to two more than the number of edges.

Solution 4.6

hexagonal prism: $v = 12$, $e = 18$, $f = 8$,   and $12 - 18 + 8 = 2$;

truncated tetrahedron: $v = 12$, $e = 18$, $f = 8$,   and $12 - 18 + 8 = 2$;

truncated octahedron: $v = 24$, $e = 36$, $f = 14$,   and $24 - 36 + 14 = 2$.

Solution 4.7

(a) If all the faces are square, then it follows from the handshaking lemma for polyhedra that

$$4f = 2e, \quad \text{or equivalently } 2f = e.$$

If exactly d edges meet at each vertex, then it follows from the handshaking lemma for graphs that

$$dv = 2e.$$

These two results can be written as

$$f = e/2 \quad \text{and} \quad v = 2e/d.$$

Substituting these two results into Euler's polyhedron formula, we obtain

$$2e/d - e + e/2 = 2,$$

which (after dividing through by $2e$) can be rewritten as

$$1/d - 1/4 = 1/e.$$

Since $1/e > 0$, it follows that

$$1/d > 1/4$$

and so $d < 4$. This means that the only possible value of d is 3. Therefore

$$1/e = 1/3 - 1/4 = 1/12, \quad \text{so } e = 12;$$

it follows that $f = 6$ and $v = 8$ — this gives the cube.

(b) Imitating the above proof with $4f = 2e$ replaced by $5f = 2e$, we deduce that

$$1/d - 3/10 = 1/e.$$

Since $1/e > 0$, it follows that $1/d > 3/10$, and so $d < 10/3$. This means that the only possible value of d is 3. Therefore

$$1/e = 1/3 - 3/10 = 1/30, \quad \text{so } e = 30;$$

it follows that $f = 12$ and $v = 20$ — this gives the dodecahedron.

Solution 5.1

Ampersand is compared with:

Bastasneeze	only by Alex;
Calamitine	only by Gill;
Dripnomore	only by Alex;
Eezybreathe	only by Eric;
Fantastamine	only by Eric;
Gladeyes	only by Gill.

Solution 5.2

- Taking the edges as the 'points' and the vertices as the 'lines', we have a regular incidence structure S_4 with $n_p = 12$, $n_l = 8$, $d_p = 2$, $d_l = 3$.

- Taking the edges as the 'points' and the faces as the 'lines', we have a regular incidence structure S_5 with $n_p = 12$, $n_l = 6$, $d_p = 2$, $d_l = 4$.

- Taking the faces as the 'points' and the vertices as the 'lines', we have a regular incidence structure S_6 with $n_p = 6$, $n_l = 8$, $d_p = 4$, $d_l = 3$.

Solution 5.3

	n_p	d_p	n_l	d_l
S_1	8	3	12	2
S_2	6	4	12	2
S_3	8	3	6	4
S_4	12	2	8	3
S_5	12	2	6	4
S_6	6	4	8	3

In each case, $n_p d_p = 24 = n_l d_l$.

Solution 5.4

S_2 and S_5 are duals, as are S_3 and S_6.

Solution 5.5

$$
\begin{array}{c}
\begin{array}{cccccc} 1 & 2 & 3 & 4 & 5 & 6 \end{array} \\
\begin{array}{c} 1 \\ 2 \\ 3 \\ 4 \\ 5 \\ 6 \\ 7 \\ 8 \end{array}
\begin{bmatrix}
1 & 0 & 1 & 0 & 1 & 0 \\
1 & 0 & 0 & 1 & 1 & 0 \\
1 & 0 & 0 & 1 & 0 & 1 \\
1 & 0 & 1 & 0 & 0 & 1 \\
0 & 1 & 1 & 0 & 1 & 0 \\
0 & 1 & 0 & 1 & 1 & 0 \\
0 & 1 & 0 & 1 & 0 & 1 \\
0 & 1 & 1 & 0 & 0 & 1
\end{bmatrix} \\
\text{(a)}
\end{array}
$$

$$
\begin{array}{c}
\begin{array}{cccccccc} 1 & 2 & 3 & 4 & 5 & 6 & 7 & 8 \end{array} \\
\begin{array}{c} 1 \\ 2 \\ 3 \\ 4 \\ 5 \\ 6 \end{array}
\begin{bmatrix}
1 & 1 & 1 & 1 & 0 & 0 & 0 & 0 \\
0 & 0 & 0 & 0 & 1 & 1 & 1 & 1 \\
1 & 0 & 0 & 1 & 1 & 0 & 0 & 1 \\
0 & 1 & 1 & 0 & 0 & 1 & 1 & 0 \\
1 & 1 & 0 & 0 & 1 & 1 & 0 & 0 \\
0 & 0 & 1 & 1 & 0 & 0 & 1 & 1
\end{bmatrix} \\
\text{(b)}
\end{array}
$$

(c) The matrix in part (b) is the *transpose* of the matrix in part (a), obtained by interchanging the rows and columns; that is, the element in row i and column j of the first matrix is the same as the element in row j and column i of the second matrix.

Solution 5.6

(a) If each 'point' degree is d_p, then the sum of the 'point' degrees must be $n_p d_p$. Similarly, the sum of the 'line' degrees is $n_l d_l$. So, from Theorem 5.1(b), $n_p d_p = n_l d_l$.

(b) A graph is an incidence structure in which the 'points' are the vertices and the 'lines' are the edges. Since each 'line' degree is 2, the sum of the 'line' degrees is twice the number of edges — which, by Theorem 5.1(b), is the sum of the 'point' (vertex) degrees. Therefore, the sum of the vertex degrees is twice the number of edges, which is the handshaking lemma for graphs.

Similarly, a polyhedron is an incidence structure in which the 'points' are the faces and the 'lines' are the edges. Exactly as for graphs, we can use Theorem 5.1(b) to deduce that the sum of the 'point' (face) degrees is twice the number of edges, which is the handshaking lemma for polyhedra.

Solution 5.7

(a) The first drug, *Ampersand* (denoted by A) is tested by Alex, Eric and Gill (denoted by $\mathbb{A}$, $\mathbb{E}$ and $\mathbb{G}$), so the first column of the block table contains the letters $\mathbb{A}, \mathbb{E}, \mathbb{G}$. The full block table is:

A	B	C	D	E	F	G
$\mathbb{A}$	$\mathbb{A}$	$\mathbb{B}$	$\mathbb{A}$	$\mathbb{B}$	$\mathbb{C}$	$\mathbb{D}$
$\mathbb{E}$	$\mathbb{B}$	$\mathbb{C}$	$\mathbb{C}$	$\mathbb{D}$	$\mathbb{E}$	$\mathbb{F}$
$\mathbb{G}$	$\mathbb{F}$	$\mathbb{G}$	$\mathbb{D}$	$\mathbb{E}$	$\mathbb{F}$	$\mathbb{G}$

The order of the open-face letters in each column does not matter.

(b)

1	2	3	4	5	6
1	5	1	2	1	3
2	6	4	3	2	4
3	7	5	6	5	7
4	8	8	7	6	8

The order of the numbers in each column does not matter.

Solution 5.8

There are many possible solutions — for example, assign the following numbers to the letters:

$$Y = 1,\ E = 2,\ W = 3,\ A = 4,\ T = 5,\ R = 6,\ H = 7.$$

Solution 5.9

The three 'lines' 1, 5 and 7 incident with the 'point' 1 are also incident with all the other 'points', once each. Thus, for each pair of distinct 'points' including 1, there is exactly one 'line' incident with both. A similar argument applies to the three 'lines' incident with any of the other six 'points', and so property (a) holds.

The three 'points' 1, 2, and 4 incident with 'line' 1 are also incident with all the other 'lines', once each. Thus, for each pair of distinct 'lines' including 1, there is exactly one 'point' incident with both. A similar argument applies to the three 'points' incident with any of the other six 'lines', and so property (b) holds.

Solution 5.10

The players in each team must be allocated numbers 1 to 7. Then, if open-face numbers represent team A and ordinary numbers represent team B, each row of the following block table represents one round of the tournament.

$$
\begin{array}{c}
\begin{array}{ccccccc}
1 & 2 & 3 & 4 & 5 & 6 & 7
\end{array} \leftarrow \text{players in team A} \\[4pt]
\begin{array}{lccccccc}
\text{round 1} \rightarrow & 1 & 2 & 3 & 4 & 5 & 6 & 7 \\
\text{round 2} \rightarrow & 2 & 3 & 4 & 5 & 6 & 7 & 1 \\
\text{round 3} \rightarrow & 4 & 5 & 6 & 7 & 1 & 2 & 3
\end{array} \left.\rule{0pt}{22pt}\right\} \text{players in team B}
\end{array}
$$

Thus:

in round 1	player 1 in team A plays player 1 in team B;
	player 2 in team A plays player 2 in team B; etc.
in round 2	player 1 in team A plays player 2 in team B;
	player 2 in team A plays player 3 in team B; etc.
in round 3	player 1 in team A plays player 4 in team B;
	player 2 in team A plays player 5 in team B; etc.

After the tournament, any pair of players from either team will be in a position to compare notes, because there will have been a player from the other team that they have both played.

Index